Chemical Research— 2000 and Beyond

Challenges and Visions

Paul Barkan, Editor
Westchester Community College

Developed from a symposium sponsored by the
New York Section of the American Chemical Society,
at The Rockefeller University,
New York, NY, October 18, 1997

American Chemical Society
Washington, DC

Oxford University Press
New York   Oxford

Chem
QD
40
.C4315
1998

Library of Congress Cataloging-in-Publication Data

Chemical research--2000 and beyond : challenges and visions / Paul
    Barkan, editor.
        p.   cm.
    Proceedings of a symposium held Oct. 18, 1997 at The Rockefeller
University.
    Includes bibliographical references and index.
    ISBN 0-8412-3575-9 (cloth)
    1. Chemistry--Research--Congresses.   I. Barkan, Paul, 1934-
QD40.C4315   1998
540' .72--dc21   98-4293
                                                                     CIP

This book was printed from camera-ready copy produced by the editor.

Copyright © 1998 American Chemical Society

Distributed by Oxford University Press

PRINTED IN THE UNITED STATES OF AMERICA

This volume is dedicated to my grandchildren, Shachar and Haim, and all the children of the world in the hope that they will have the opportunity to experience the excitement and thrill of exploration and discovery as did the contributors to these proceedings in a world committed to the support of basic scientific research for the benefit of humankind.

Paul Barkan

The American Chemical Society's New York Section
presents

Challenges and Visions:
Chemical Research—2000 and Beyond

Chemical scientists join Nobel laureates, and industry and government
representatives in a dialogue for the future

Saturday, October 18, 1997
The Rockefeller University, New York, NY

Cosponsored by The Rockefeller University

Conference Participants

Dr. Torsten N. Wiesel
President, The Rockefeller University
Nobel Prize in Physiology or Medicine, 1981

Prof. Paul Barkan
Chair, New York Section, American Chemical Society

Dr. Mario J. Molina
Lee and Geraldine Martin Professor of Environmental Sciences,
Massachusetts Institute of Technology
Nobel Prize in Chemistry, 1995

Dr. George A. Olah
Donald P. and Katherine B. Loker Distinguished Professor of Organic Chemistry,
University of Southern California
Director, Loker Hydrocarbon Research Institute
Nobel Prize in Chemistry, 1994

Dr. Paul S. Anderson
Senior Vice President of Chemical and Physical Sciences,
The DuPont Merck Pharmaceutical Company
President, American Chemical Society, 1997

Dr. Dudley R. Herschbach
Frank B. Baird, Jr. Professor of Science,
Harvard University
Nobel Prize in Chemistry, 1986

Dr. William N. Lipscomb, Jr.
Abbott and James Lawrence Professor, Emeritus,
Harvard University
Nobel Prize in Chemistry, 1976

Dr. Robert F. Curl
Harry C. and Olga K. Wiess Professor of Natural Sciences
Chemistry Department and Rice Quantum Institute
Rice University
Nobel Prize in Chemistry, 1996

Dr. Ronald Breslow
Samuel Latham Mitchill Professor of Chemistry and
University Professor, Columbia University;
Immediate Past President, American Chemical Society

Hon. Robert S. Walker
President, The Wexler Group
U.S. House of Representatives (Retired), Chairman, House Science Committee,
Chairman, House Republican Leadership

Robert F. Service
Research News Writer, AAAS *Science*

Invited Contributors to the Proceedings

Dr. W. O. Baker
Retired Chairman of the Board
Bell Labs Innovations; Lucent Technologies

Dr. Alan J. Main
Senior Vice President, Research
Novartis Pharmaceuticals Corporation

Dr. K. C. Nicolaou
Aline W. and L. S. Skaggs Professor of Chemical Biology, and Darlene Shiley
Chair in Chemistry: Chairman, Department of Chemistry;
The Scripps Research Institute
Professor of Chemistry, University of California, San Diego

Dr. Francis A. Via
Director of Contract Research, Corporate Programs
Akzo Nobel

Dr. Brandon H. Wiers
Manager, External Research Programs (Retired)
Proctor & Gamble Co.

Contents

Knowledge-Driven Research: New Tools, New Frontiers

The Chemical Industry: Research for Competitiveness

Government and Media: Support for the Research Endeavor

Chemistry: The Central, Creative, and Enabling Science
Comments and Discussion

Preface

These proceedings represent the commitment of the American Chemical Society's New York Section to continue the dialogue on the future directions of chemical research initiated through our symposium "Challenges and Visions: Chemical Research—2000 and Beyond" at The Rockefeller University in New York City on October 18, 1997. That event, cosponsored by The Rockefeller University, addressed the effects of the global political, economic and social changes which are threatening the pace of progress through scientific research and stressed the urgency for the chemical community to assume an active role in convincing policy makers and the public that the quality of life in the 21st century will depend on a strong national science agenda that fosters basic scientific research. This agenda must recognize that chemistry, through its impact on the other scientific disciplines, has had and will continue to have a critical role in the health, welfare, and economic growth of our nation.

At the symposium the need for governmental and industrial support for basic research was underscored by the five Nobel Laureates in chemistry who presented their research and its implications for the future society. They were joined by distinguished chemists from academia and industry, legislators involved in science policy, and media representatives who offered their perspectives on the conditions necessary for our nation to maintain a leadership research environment. The concluding panel discussion assessed some of the problems facing the research community and presented recommendations for maintaining the pace of scientific research.

This volume comprises the edited lectures of the ten presenters and five invited contributions and concludes with a selection of remarks made by the presenters and the audience during the panel discussion. It was not possible, in a one-day symposium, to include all the important areas of chemical research or other factors, such as reform of chemical education from elementary through graduate school and protection of intellectual property, which will effect the vitality of research in the next century. It was hoped, rather, that by highlighting some areas of chemical research which may provide solutions for future global problems, we would be able to contribute to the input being provided to Congress as it struggles to formulate a national science policy. Through these proceedings, the New York Section of the American Chemical Society hopes to encourage additional forums throughout the chemical community nationwide. This is in keeping with the consensus of the 370 attendees at the symposium, comprising chemists from academia, industry and government laboratories, graduate and undergraduate students, high school teachers and students, patent attorneys, and representatives from the investment community.

The volume opens with the introduction to the symposium expressed in a distressing figurative metaphor alluding to the deterioration of public support for scientific research and underscoring the dangers of an inadequate response to the erosion of this support. In these remarks, the scientific community is exhorted to become advisors, advocates, and activists in order to ensure the progress of science and its benefits to society.

The second section discusses the role of chemical research in providing solutions to critical problems facing humankind in the next century: namely, the forecasted doubling of the world's population by the year 2050, the depletion of fossil fuels, our major source of energy, and the need to protect the integrity of the Earth's atmosphere from the consequences of human activities. George Olah raises an urgent call for a commitment to the basic chemical research required to convert the greenhouse gases CO_2 and H_2O to methanol through an electrocatalytic process, and for the development of nuclear energy as a safe and efficient energy source. Mario Molina cites the need for a vigorous research program involving regional and global atmospheric chemistry and the importance of the public's understanding of its consequences in order to construct a rational program to protect our planet. Robert Curl uses his research in the field of combustion chemistry and the development of portable infrared monitors for atmospheric gases to send a powerful message that the problems common to all humanity—energy production, sufficient food to meet the needs of a burgeoning population, and the abatement of air pollution—will require a renewed commitment to support basic scientific research.

The need for a national science policy that recognizes the importance of curiosity and idea-driven research as the real seeds of new science is the focus of the third section beginning with William Lipscomb's tracing of the development of the relationship between molecular structures and their function, including those related to living systems. K. C. Nicolaou then reflects upon nature's creative secrets through the power of chemical synthesis. Using the work of his own laboratory, he explores the future of organic synthesis, as the enabling technology for biology and medicine, which will enhance our understanding of life and lead to the rational design of new medicines. The application of such research in the pharmaceutical industry is described in Alan Main's article on the development of two new drug discovery paradigms: "standardize the product, where a single chemical class of compounds is being developed as potential universal therapeutic agents; and standardize the process, where the drug discovery process itself is being highly optimized to efficiently deliver new therapies."

Knowledge-driven research not only produces new concepts but allows for the further exploration of existing ones. Dudley Herschbach, in section four, discusses the trapping of molecules and the liberating of catalysts to highlight how the development of new tools has opened exciting vistas of research in physical chemistry.

To bring the benefits of basic chemical research to our society requires a commitment of all sectors. Using the recently formulated long-range plan of the chemical industry, *Technology Vision 20-20: The U.S. Chemical Industry,* Paul Anderson presents the role of the chemical industry in the emerging global setting and the need for academe, government and industry to work in consonance with shared goals. This focus in section five continues in Francis Via's article defining the critical factors that drive today's chemical industry and his belief that industry's global position will be strengthened by partnerships teaming the three sectors: government, industry, and the university. Brandon Wiers recommends amending the Research & Experimentation Tax Credit to promote private sector investment for basic scientific research by universities and government laboratories.

A sustainable research policy will require knowledgeable policy makers and a scientifically literate public. In the sixth section, Congressman Robert Walker calls on the scientific community to become advocates for support of research by convincing policy makers of the critical need for funding through budget authorizations to assure long-term sustainability of their research. Adequate support will require seeking innovative ways to obtain funding, including international consortia and changes in tax policy. Science research writer Robert Service explores the reasons for the declining media coverage of science topics, especially chemistry. His recommendations to the scientific community on how to promote coverage includes workshops for media people.

In the concluding articles, Ronald Breslow reviews the strength of chemistry as the central, enabling, and creative science, and describes its contributions to human welfare. He predicts that chemistry will meet the challenge of the future in the areas of: health, life, the environment, materials, computers, electronics, catalysis, synthesis, structure determination, and reaction mechanism. Chemistry, having reached a remarkable level of sophistication in elucidating the structure of matter, is now ready, according to the commentary by W. O. Baker, to assume its place in our educational culture alongside physics, astronomy, and cosmology. Thus, children at an early stage would be able to experience that everything they handle is an aggregate of chemicals capable of being defined. Such an approach would then become an inherent part of the educational development of our society and could serve to energize chemical research. The urgent need for initiating W. O. Baker's recommendation is underscored in a recent report of the Amsterdam-based International Association for the Evaluation of Educational Achievement, according to which American high school seniors are among the industrial world's least prepared in mathematics and science. Particularly disappointing in the report entitled the *Third International Mathematics and Science Study* is the finding that our most advanced students in mathematics and physics ranked the lowest among the twenty three countries participating in the study.

The New York Section extends its gratitude to The Rockefeller University for its cosponsorship of the symposium which made it possible to initiate our program at its world-renowned center for scientific research and graduate education. Its impressive setting honored our distinguished presenters. The commitment, vision, and generosity of ten corporations and two private foundations brought "Challenges and Visions: Chemical Research—2000 and Beyond" to the community of concerned individuals and will now provide for the input of the scientific community through the preparation and dissemination of this volume to federal government lawmakers, corporations, universities, professional scientific societies, the media, and the public.

We wish to express our appreciation to the ACS Books Department, and, in particular, to Cheryl Wurzbacher, Production Manager, for her guidance throughout the preparation of this volume and to Elaine Galen for her artistic interpretation of Logos and Mythos in Chapter One.

My deepest appreciation to my wife, Dr. Roberta Barkan, for her countless hours of dedicated proofreading of the manuscripts and especially for her unique talent of keeping my spirits high during moments of frustration with this project.

The contributing authors express the wish that the ideas contained in this volume will generate further discussions through symposia and workshops, and will help to convince policy makers and the public that the future of our global society is inextricably linked to a strong basic science research agenda. As chemical scientists, we have an obligation to be involved in assuring that the resources and opportunities will be available for future generations to continue in the pursuit of scientific knowledge and its concomitant benefits to society. We ask you, the reader, to join us in this vital endeavor.

Paul Barkan
Symposium Organizer & Chair
27 Wynmor Rd.
Scarsdale, NY 10583
February 1998

Acknowledgments

The American Chemical Society's New York Section gratefully acknowledges the generous support which has made possible the symposium and the proceedings.

Leadership Challenge Grant	The Smart Family Foundation
Benefactors	The Camille and Henry Dreyfus Foundation Special Grant Program in the Chemical Sciences Novartis Pharmaceuticals Corporation Pfizer Central Research Division
Patron	Merck Research Laboratories
Sponsors	Akzo Nobel Boehringer Ingelheim Pharmaceuticals Polychrome Corporation Wyeth-Ayerst Research
Supporters	Bristol-Myers Squibb Pharmaceutical Research Institute Melles Griot Schering-Plough Research Institute

The Chemical Community: Inaction versus Action

1

Myth Versus Reality: Huizinga Reinterpreted

Paul Barkan

1997 Chair, New York Section, American Chemical Society
Westchester Community College, Chemistry Department
Valhalla, NY 10595

Challenges and Visions: Chemical Research—2000 and Beyond was convened by the American Chemical Society's New York Section as a response to the critical issues which will affect the pace of scientific research in the 21st century. This introduction to the Symposium addresses the need for the chemical community to become Advisors, Advocates and Activists to ensure the progress of science and the concomitant benefits to the economic and social well-being of society. The current situation is described in figurative terms underscoring the dangers of an inadequate response to the erosion of support for basic research.

Barbarisation sets in when, in an old culture, which once, in the course of many centuries, had raised itself to purity and clarity of thought and understanding, the vapours of the magic and fantastic rise up again from the seething brew of passions to cloud the understanding: when the muthos *supplants the* logos.

J. Huizinga, *In the Shadow of Tomorrow.* *(1)*

When Logos Triumphed

Not so long long ago, high on Mt. Logos, in the land of ASU, scientists were engaged in the pursuit of knowledge; and in all of ASU there thrived a highly developed civilization. Asuvians believed in progress and opportunity. Their seers assured them that their continued advancement would depend on the strength of their science and technology. Their Elected Leader even proclaimed "we must not shrink from exploring the frontiers of science. Our scientists have blended the elements of earth, air, fire and water and have created materials for our comfort and well-being." And so, Asuvians supported basic research, invention and creativity. They revered and lauded education and the accomplishments of their scientists. Indeed, this was a Golden Age for science and discovery. And the ASU population enjoyed a comfortable standard of living, secure in the belief that improved health, longevity and prosperity would continue to advance and that their frontiers would remain secure from foreign invaders.

So, scientists were congregated in the Elysian Fields of Mt. Logos where they could devote all their energies to their quest to unravel the mysteries of the universe. They enjoyed the respect of the people and the support of their elected councils. There were limitless opportunities for research in their universities and industrial laboratories. They benefited from the boundless assistance of masses of devoted students and were duly credited with enhancing the quality of life in ASU.

Now as scientists, they stood apart from the rest of ASU society. They spoke in varieties of dialects characterized by symbols and formulas that seemed great mysteries and foreign to the other Asuvians. They created, explored and dreamed, and Mt. Logos seemed ever taller and higher. Absorbed in their research, they had no need to be involved in the issues before their councils or in explaining their objectives to the masses of Asuvians. They were, indeed, left alone in their pursuits.

Days passed and became years. A new century loomed ahead. The knowledge of ASU began to spread to other mountains and valleys, as slumbering inhabitants awoke to the opportunities. They, too, began to foster their own scientists. (Fig. 1)

The Awakening of Mythos

At the same time, long smoldering animosities between peoples diminished and boundaries disappeared. Technologies began to bring peoples, industries and universities closer and knowledge was soon transported across the newly drawn frontiers. New horizons loomed for scientists beyond Mt. Logos and science was glorified beyond its Elysian fields. Gradually, a new tellurian society began to emerge in competition with the singular supremacy of ASU. This, too, at first seemed good. Then, slowly but soon in increasing numbers, shadowy apparitions began to rise. (Fig. 2)

Long suppressed in the bowels of Mythos, they now found media receptive to their cries. Bellowing in rage, they mocked, taunted and screamed: "A plague upon the Elysian scientists, no more should they enjoy unbridled support. They have already discovered all that is important, their efforts are but vanity and they are venturing dangerously into the domain of the gods." These Specters even conjured up strange numbers to claim that there was no correlation between their council's spending on science and technology and Asuvian prosperity. They cursed the benefits the scientists had wrought claiming that the legacy was but Asu's murky lakes and rivers and fouled air.

Now many Asuvians were drawn to these dark visions and revived their beliefs in ghosts and the supernatural. They began to question established scientific truths and blamed the spirits of Science and Technology for the diverse problems in their society. They even petitioned their councils to withhold support for the Elysian Fields.

How easily were the voices of these Shades seized upon by ruling councils seeking to balance the stretched Asuvian budget. Years of spending on defense, social

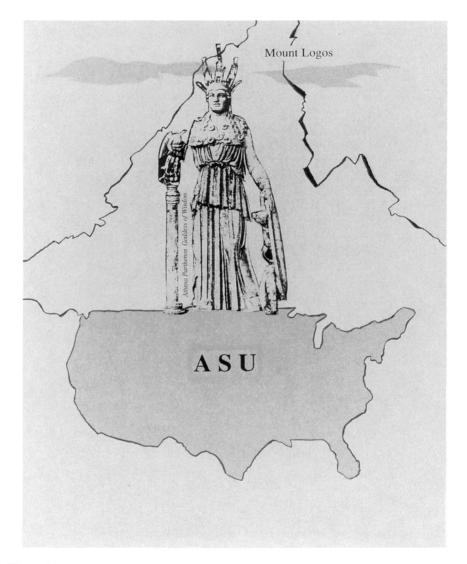

Figure 1.

Figure 2.

welfare programs, education, science, and support of friendly neighbors, had taken its toll on the Asuvian treasury. The debt had reached astronomical proportions. Stringent measures were to be applied. No longer could funds be spent on pure research; the immediate needs of the ASU society would have to be served. Industry, too, felt the need to restructure in order to maintain its position in the new society.

Suddenly, so suddenly it seemed, the glorious opportunities for pure scientific exploration in ASU could come to an end and scientists would begin to know the pain of abandonment and isolation. The long range prospects for continued advancement were dimmed. No longer would ASU be a supreme entity in the emerging tellurian society.

From Myth to Reality: A Cause for Chemists

Now myths by their very nature must glorify the good and demonize our nightmares. And the tale just told is merely a simplification of complex events and diverse causes that threaten the future of basic research as we approach the next century. The narrative presents but the shells of large scale issues that demand our serious attention.

As chemists, we have chosen our careers out of dedication to our science, lured by the excitement of discovery and the capacity to contribute to the betterment of the quality of life. We believe that chemistry, as the central science through its impact on biology, medicine and technology, has been and should continue to be essential to the growth and prosperity of our nation. Yet, we are all keenly aware of the realities and policies that are challenging this vision and endangering the ways in which science has been practiced, taught, and ultimately perceived by the public.

Our leadership in science and technology is being threatened by the rapidly emerging global industrial competition, the staggering merchandise trade deficit, the loss of dominance in some critical technologies, short-sighted policies with regard to our future energy needs, and pressures on corporations to satisfy stock holder demands for rapid returns on their investments. Increasing science illiteracy at all levels of the population and the rising influence of deconstructionists debunking established scientific truths encourage legislators who now seem determined to balance the budget at the expense of scientific research and progress. The effect of these forces will be a slowing in the dramatic rate of scientific advances that marked this century's "golden era" of research. Unless the trends are reversed, we will witness the erosion of the social, cultural and economic infrastructure of our society. (2)

As chemists, therefore, we have an obligation to become involved in issues that affect the progress of science and the future of our profession. Individually and collectively, through our professional scientific societies, there are measures we can take and activities in which we can participate. For example, it is both appropriate and essential that chemists establish a relationship with their elected legislators by offering to advise them on issues of chemical science that relate to public policy. More of us need to become knowledgeable about the federal budgetary process and its impact on the nation as a whole. From this vantage point, we may be able to impress upon our legislators that the quality of life in our future society requires support for basic research and science education.

It is equally important that this message reach the public through education and through the media. As chemists we should take an active interest in our local school districts' policies on science education and offer our support and advice in areas of our expertise. This is one way to ensure that the science that is taught is based on established scientific principles. It is often possible for chemists to assist teachers in the classroom through hands-on activities, particularly through the vehicle of National Chemistry Week as well as other science recognition days throughout the calendar. Local ACS sections and National ACS have the expertise and support materials to assist in the planning of such events.

Since public perceptions of science are influenced by presentations by the media, it is necessary to educate the media to the importance of chemistry. It is ironic that chemistry, which has contributed and will continue to contribute to society as the creative science that is central to the other scientific disciplines, is dismissed by the media as being obsolete. Examples of this abound in the lack of coverage of chemical advances in the press. An interesting item in a recent issue of *Science Times* (*3*) is to the point. In reporting the results of a survey taken on the WWW in which scientists were asked "What is the question you are asking yourself?", of the 49 scientists whose response was included, one was a biochemist. No other chemist was cited. We must find the appropriate means to educate those who report on science to ensure accurate coverage of chemistry and science. Only this way can we expect to find fair representation of those areas of discovery which are truly significant. The media presents what it perceives to be of interest to the public, and, therefore, we must learn to present our science in terms that are understandable to the layman and which convey the excitement and the importance of research.

In 1966, in a lecture before a group of science teachers, Linus Pauling set forth his social credo for scientists which began with a statement germane to chemists today.

> We, as scientists, have the general social responsibilities
> resulting from our knowledge and understanding of science
> and its relation to the problems of society. It is not our duty to
> make the decisions, to run the world. It is, rather, our duty to
> help educate our fellow citizens, to give the benefit of our
> special knowledge and understanding, and then to join with
> them in the exercise of the democratic process. (*4*)

Therefore, as chemical scientists, we must become Advisers, Advocates and Activists and bring our knowledge, expertise and concerns to the forefront of discussion that will affect support for basic research. If we fail to act, Huizinga's specter of the barbarization of knowledge—of logos—will become a reality.

Literature Cited

1. Huizinga, J. *In the Shadow of Tomorrow;* W.W. Norton & Company, Inc.: New York, NY; 1936; p 216.
2. Barkan, P. *The Indicator.* December 1996, p 4.
3. *Science Times, the New York Times*, December 30, 1997, p 3.
4 Marinacci, B., Ed.; *Linus Pauling in His Own Words;* Simon & Schuster Inc. : New York, NY; 1995; p 190.

Global Crisis: Population and Energy

2

The Changing Chemistry of the Atmosphere: A Challenge for the 21st Century

Mario J. Molina

Department of Earth, Atmospheric, and Planetary Sciences and Department of Chemistry, Massachusetts Institute of Technology
77 Massachusetts Avenue, Cambridge, MA 02139-4307

Atmospheric chemistry is a relatively new field which has become very important for assessing the extent to which the Earth's atmosphere is being changed as a consequence of human activities. The chemistry of the stratosphere is now reasonably well established, although additional research is needed in areas such as the formation mechanisms and chemical properties of polar stratospheric clouds. The chemistry of the troposphere is more complicated; much remains to be learned in order to better understand at a fundamental, molecular level processes such as the degradation of hydrocarbons, and heterogeneous reactions involving tropospheric aerosols.

I would like to present a brief overview of the type of research I have been involved with in the past years, setting the stage to make some remarks about where this research is going. My work involves atmospheric chemistry; so, let me remind you how the atmosphere functions. What I would like to do is to specifically mention some examples of stratospheric chemistry, and then contrast it with the chemistry occurring closer to the earth's surface.

Structure of the Atmosphere

Figure 1 shows a typical temperature profile, temperature varying with latitude and with season. The way the atmosphere functions is that temperature drops with altitude in the lowest layer, the troposphere, but then it increases, leading to a so-called "inverted temperature profile", which is what characterizes the next layer, the stratosphere. We all know that as you climb to higher altitudes it gets colder — there

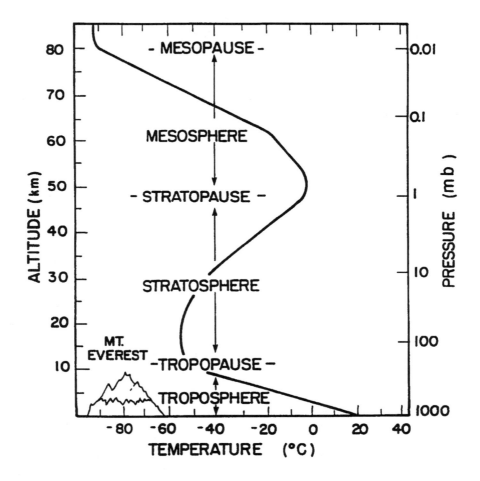

Figure 1. Typical atmospheric temperature profile

is always snow at the top of high mountains. An important concept is the way this profile affects atmospheric motions: mixing is quite rapid in the lowest layer; in contrast mixing, particularly in the vertical direction, is very slow in the stratosphere. So, it takes quite a bit of time for chemical species released at the earth's surface to reach the stratosphere.

But there is another very important distinction between these two layers. In the troposphere there are clouds and rain, which provide a very efficient mechanism to remove most natural or industrial pollutants emitted at the surface. There are also some oxidation processes that remove species that are not water soluble, such as hydrocarbons. As a consequence, air entering the stratosphere is relatively clean, and its chemistry is simpler than that in the troposphere. Because of this circumstance, we were able to take a reductionist approach in stratospheric chemistry: we can actually sort out all the major species that we believe are important in terms of their chemistry, and we can characterize all the reactions that inter-convert these species among themselves.

Stratospheric Chemistry

There are three types of species in the stratosphere (see Figure 2): the first one is sources, which are released at the earth's surface. Some of those are industrial, like the CFCs, and the others are of natural origin, but all of them are very stable species. Except for water itself, they are emitted in reduced form, but are oxidized in the atmosphere, eventually producing water-soluble products. But in reduced form they are not water soluble, so they are not removed efficiently in the troposphere. Consequently, air currents can move the source species into the stratosphere, which is where they are destroyed and where they generate free radicals.

Free radicals are chemically are very reactive: they have an unpaired electron. In fact, all of the atmospheric free radicals shown in Figure 2 have an odd number of electrons. These are the species that are involved in catalytic cycles that can affect the ozone balance. But the free radicals may also react with each other or with the source species, producing compounds that are called "temporary reservoirs". They are temporary because after subsequent chemical or photochemical reactions in the stratosphere they release again free radicals. The situation then is that whatever reaches the stratosphere in the form of a source species, eventually gets down to the lowest layer — the troposphere — in a water-soluble form that is very rapidly removed by rain.

So, much of the research in atmospheric chemistry, as far as the stratosphere is concerned, involves measuring the chemical and photochemical reaction rates for inter-conversion of all these species, making it a very rich research topic. The approach that we were able to take is essentially a molecular point of view, investigating at each elementary chemical reaction.

Atmospheric chemistry is very much an experimental science. There is a very well developed theoretical framework which helps us understand the kinetics and the mechanisms of atmospheric reactions, but theory is not yet at the stage where you can actually predict the reaction rates reliably from first principles (either for gas phase

Trace Species in the Stratosphere		
Sources	Radicals	Sinks
N_2O	NO, NO_2	HNO_3 HO_2NO_2 $ClONO_2$
H_2O CH_4	OH, HO_2	H_2O H_2O_2 HNO_3
$CFCl_3$ CF_2Cl_2 CH_3Cl	Cl, ClO	HCl HOCl $ClONO_2$
CF_3Br CF_2ClBr CH_3Br	Br, BrO	HBr HOBr $BrONO_2$

Figure 2. Classes of stratospheric trace species

reactions or for photochemical reactions). So, one has to carry out experiments to measure the rates of all these processes. Another very important activity consists of measuring the concentration of the various species in the atmosphere, to check our understanding of the system.

The Antarctic Ozone Hole

Figure 3 shows the results of satellite measurements, yielding total amount of ozone, integrated from the earth's surface to the top of the atmosphere (*1*). The figure shows that over Antarctica -- where the stratosphere gets very cold — ozone levels drop very dramatically in the Spring months; that is, what, we know as the Antarctic ozone hole. Ozone is predominately made in the tropics, but significant amounts move to high latitudes. We have, fortunately, measurements going back to the 60's, and so, it is clear that this particular phenomenon — the ozone hole — started developing only recently, in the 80's and 90's.

Figure 3 also shows satellite measurements of chlorine monoxide, the free radical derived from the CFCs. The concentrations of this free radical in the polar vortex are so large that they can indeed be measured remotely, and that's rather unusual. Free radicals are usually too reactive to be present in large enough abundance to be easily measurable. But, in this case it is clear that this free radical is so abundant over the poles that it can be monitored by the satellite through its emission of microwave radiation.

Figure 4 is another view of the Antarctic ozone hole: it shows balloon measurements of ozone conducted at the South Pole (*2*). The figure shows that in the lower stratosphere, over an altitude range of about 5 km, more than 99% of the ozone disappears in a matter of weeks.

The Montreal Protocol

Because the scientific case was so strong for this issue, it was possible to get all the industrialized countries, as well as many developing countries, to agree to do something about the problem: they formulated an international agreement, the Montreal Protocol. It calls for the complete phase out of CFCs by the end of 1995 in industrial countries; developing countries are allowed to continue producing these compounds for a few decades, in order to have a smooth transition to CFC free technologies. Atmospheric measurements are already showing that the global concentrations of CFCs are beginning to decline as a consequence of the Montreal Protocol.

However, atmospheric measurements of the Halons do not show yet that their concentrations are decreasing, even though they are no longer being manufactured in industrial countries because they are also controlled under the Montreal Protocol. The Halons are industrial compounds used as fire extinguishers; they are chemically similar to the CFCs, but contain bromine, in addition to chlorine. Bromine turns out to be much more efficient than chlorine in terms of ozone destruction because the temporary reservoirs for bromine are less stable than those for chlorine, and hence bromine remains as a free radical a larger fraction of the time. But developing

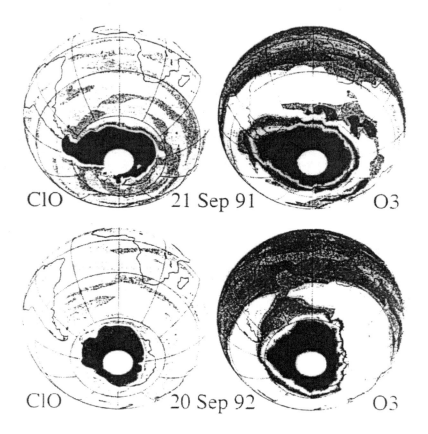

Figure 3. Measurements of column abundances of the C10 radical and of ozone
above ~16 km in September 1991 and 1992, carried out by the UARS MLS
satellite instrument. The dark regions represent a column abundance of
C10 larger than 10^{19} molecule m^{-2}, and a column abundance of ozone
smaller than 185 Dobson units. Adapted from (1).

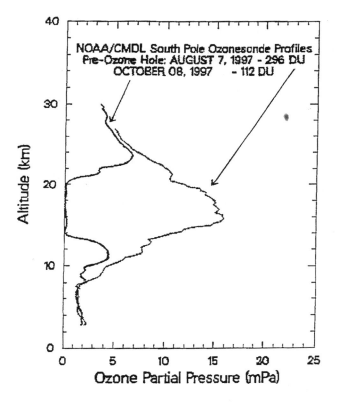

Figure 4.   Ozone profiles measured in 1997 over the South Pole with the balloon instrument described in (2).

countries such as China appear to be still manufacturing and releasing significant amounts of Halons; this is one remaining problem that needs to be addressed with some urgency.

Laboratory Measurements

Let me describe now some laboratory experiments that we have carried out in the past. One of them involves the chemistry of gas phase species on the surfaces of cloud particles. Over Antarctica, some of those cloud particles are made of water ice. We and others (3,4) were able to show that these clouds play a very important role in promoting the conversion of chlorine reservoir species to molecular chlorine, which absorbs light very readily to yield free chlorine atoms. The important property that distinguishes molecular chlorine from other temporary reservoirs is that it is photolytically very active; it is a green gas, and hence absorbs visible radiation.

It was a surprise that ice at -80°C behaves in such a way that hydrogen chloride, HCl, actually dissolves and ionizes on its surface, as it does on liquid water. It appears that the surface of ice is rather disordered, so in the presence of HCl it behaves like liquid water, even at very low temperatures (5). This explains how is it that trace amounts of HCl vapor react very efficiently with hypoclorous acid or with chlorine nitrate in the presence of ice, releasing molecular chlorine.

Let me describe next another experiment that we have conducted recently in our laboratory. It is the measurement of the rate of reaction between two free radicals: OH and ClO, both of which are very important in the stratosphere. Experimentally, it is very challenging to investigate this type of reactions: OH reacts very fast with itself, and ClO reacts fast with itself; we need to generate and rapidly mix both species, and we need to measure their concentrations as a function of time; etc. But we have managed to develop new techniques, using flow tubes, to conduct accurate rate constant measurements for complex systems like this one. Most of the time the OH + ClO reaction yields two other radicals, namely Cl and HO_2; however, some fraction of the time the products are HCl and O_2, and it is this channel that matters most in the atmosphere, because it yields two temporary reservoirs, that is, two species which are not free radicals. We have been able to show that these reservoirs are produced with a five to seven per cent yield (6). With this result, the agreement between model calculations of the chemistry of the upper stratosphere and atmospheric observations improves very significantly.

There are many other examples which show us that we still need to unravel at a very fundamental level the mechanism of reactions that matter in the atmosphere, in order to have confidence in our predictions of human-induced environmental changes; there is certainly much work to be done.

Photochemical Air Pollution

Another important atmospheric problem that you are all aware of occurs, for example, in Los Angeles; it is urban "smog". Mexico City, where I was born, is now extremely polluted, as are many other large cities around the world. Besides producing smog, other human activities that contribute to the pollution of the lower atmosphere are the burning of forests, and "slash and burn" agriculture, which is practiced mainly in the tropics. As we learned in recent weeks, large forest fires also occur in Southeast Asia. In terms of atmospheric pollution, biomass burning activities and urban smog have very similar effects: what is normally considered a very localized problem — urban smog — is becoming a large scale regional problem, namely pollution of the troposphere.

An important consequence of this pollution is an increase in the amount of ozone that exists close to the earth's surface. Figure 5 provides an indication that at the beginning of the century the troposphere had lower ozone levels than at present. The figure represents background ozone measurements, not localized to cities. Thus, the amounts of ozone in the entire globe close to the surface have increased. Unfortunately, this does not compensate for ozone depletion in the stratosphere, as far as ultraviolet radiation effects are concerned. High ozone levels in cities have very important consequences for human health, and on a global scale, the increase in ozone has damaging effects on a variety of ecological systems; some food crops, for example, turn out to be rather susceptible to this damage.

Let me get back to atmospheric chemistry: the basic ingredients of smog — as well as of tropospheric pollution — are nitrogen oxides, volatile organic compounds such as hydrocarbons, and solar radiation. Ironically, nitrogen oxides deplete ozone in the stratosphere, but generate it in the lower atmosphere.

The mechanism of oxidation of the simplest hydrocarbon — methane — involves about a dozen elementary steps that is, reactions representing actual molecular events; most of these steps are well characterized. This type of chemistry gets rapidly very complicated as the size of the hydrocarbon molecule increases. If you consider, for example, butane with its four carbon atoms, the number of intermediates in the oxidation -- which eventually yields water and CO_2 — is already so large that only a small fraction of the elementary steps have been characterized in the laboratory.

In terms of elementary steps we do not expect to reach the same sort of complete picture of smog chemistry as we have for the chemistry of the stratosphere, which is much simpler and for which we have studied directly practically all the important elementary chemical and photochemical reactions. But, we should have a better understanding of many more of the elementary steps, so that we can predict with confidence how the chemistry of the lower atmosphere will change as a consequence of human activities.

Conclusion

Let me present now some concluding thoughts. First of all, from a planetary perspective the atmosphere is very, very thin — most of its mass is confined to the first ten or twenty kilometers above the Earth's surface, whereas the distance between poles is 20, 000 km. The atmosphere is, thus, a rather fragile shield; it is important that we understand how it functions, so that we can better assess to what extent is it being affected by human activities. We have to continue carrying out a vigorous scientific research program involving regional and global atmospheric chemistry, and we have to communicate our findings to society at large, so that adequate steps may be taken to prevent serious damage to this thin gaseous shield which surrounds our planet, and which is such an essential component of our environment.

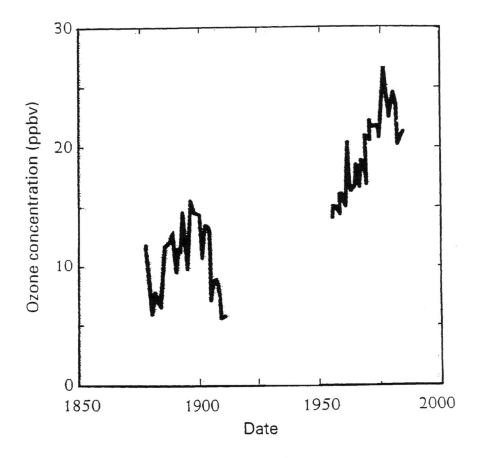

Figure 5. Surface measurements of ozone representing background tropospheric
           levels. Adapted from (7).

Literature cited

1. World Meteorological Organization, *Scientific Assessment of Ozone Depletion, 1994*; Washington, D.C.: NASA/NOAA/UKDOE/UNEP/WMO; Global Ozone Research and Monitoring Project, Rep. No. 37; 1994.
2. Hofmann, D.J.; Oltmans, S.J.; Lathrop, J.A.; Harris, J.M., Vomel, H.; Record low ozone at the South Pole in the spring of 1993. *Geophys. Res. Lett.* **1994**, *21*, pp 421-424.
3. Molina, M.J.; Zhang, R.; Wooldridge, P.J.; Mcmahon, J.R.; Kim, J.E.; Chang, H.Y.; Beyer, K.D.; "Physical-chemistry of the $H_2SO_4/HNO_3/H_2O$ System - Implications for Polar Stratospheric Clouds", *Science*, **1993**, *261*, pp 1418-1423.
4. Kolb, C.E.; Worsnop, D.R.; Zahniser, M.S.; Davidovits, P.; Keyser, L.F.; Leu, M.T.; Molina, M.J.; Hanson, D.R.; Ravishankara, A.R.; Williams, L.R.; Tolbert, M.A.; Laboratory studies of atmospheric heterogeneous chemistry, in *Current Problems and Progress in Atmospheric Chemistry*, Barker, J.R., ed.; World Scientific Publishing, 1995.
5. Molina, M.J.; The probable role of stratospheric 'ice' clouds: Heterogeneous chemistry of the ozone hole, in *Chemistry of the Atmosphere: The Impact of Global Change*, Calvert J.G., ed., Blackwell Scientific Publications, 1994.
6. Lipson, B.L.; Elrod, M.J.; Beiderhase, T.W.; Molina, L.T.; Molina, M.J.; "Temperature dependence of the rate constant and branching ratio for the OH + ClO reaction", *J. Chem. Soc., Faraday Trans.* **1997**, *93*, pp 2665-2673.
7. Volz, A.; Kley, D.; Evaluation of the Montsouris series of ozone measurements made in the nineteenth century, *Nature,* **1988**, *332*, pp 240-242.

3

Tunable Infrared Laser Sources:
From the Laboratory Towards the Rice Paddy

Robert F. Curl

Chemistry Department and Rice Quantum Institute
Rice University, Houston, TX 77005

The development of portable infrared monitors for atmospheric gases is placed in the context of the enormous problems facing humanity in the coming century. The U. S. A. will be forced to come to grips with these problems, the solution of which will require a renewed commitment to research.

We begin with a short research story. Then there is going to be a moral. A large fraction of this paper is going to be concerned with the moral.

Infrared Kinetic Spectroscopy and Infrared Laser Sources

We have been working for some years in the field of combustion chemistry. The rationale for this is that almost all the energy in the U.S.A. comes from the combustion of fossil fuels, and, therefore, it seems important to understand combustion processes. For example, if we understand combustion chemistry, then we might be able to do some practical things such as reduce pollution by changing combustion conditions or reduce knock in engines with higher compression. In fact, improvements of both these types have already been made by others and progress continues. Because of such considerations, the Department of Energy has a small program of fundamental studies of combustion chemistry. Graham Glass and I and our students are a small part of this program. Frank Tittel will come into this story: he and I collaborate in closely related work.

Combustion consists of a set of free radical chain reactions. The things of practical interest such as soot formation are part of this free radical kinetic system. Our role in the program is to find ways of observing the small radicals involved and to measure their reaction rates and determine their reaction products. The technique that we have been using is infrared kinetic spectroscopy as shown in Fig. 1. In this technique, a free radical is produced by flash photolysis of some suitable precursor. Then the transient infrared absorption of the radical is observed using a tunable infrared laser probe.

As an example, consider rate constants. The rate constant of a reaction of the radical with a stable reagent species can be measured by introducing an excess of the reagent and following the rate at which the radical disappears as a function of time. (*1*) Because a large excess of reagent is added, the reagent concentration varies very little over the course of the reaction and the radical concentration decays exponentially with time (this is called pseudo first order kinetics). The exponential decay constant is determined for a given reagent concentration by fitting the decay of the infrared absorption of the radical. By repeating this at several reagent concentrations and plotting the decay rates versus the reagent concentration, the reaction rate constant can be obtained as the slope of the resulting straight line. Rate constants can be determined by this approach from dry ice temperature to above 1000 °C.

Many radical reactions have more than one pathway. By observing the products of the radical reactions and making quantitative infrared intensity measurements, one can determine the branching into each reaction channel. (2)

The infrared kinetic spectroscopy experiment is very simple. The technologically challenging part is producing the tunable cw single frequency infrared laser probe. Consequently, most of our thinking about making the experiment work better has been concerned with the probe laser. A large fraction of our work has been done with color center lasers, which cover the higher frequency part of the mid-infrared permitting us to observe O-H, N-H and C-H stretching fundamentals. In order to access other fundamental vibrations, we searched for other tunable, single-frequency cw infrared lasers at longer infrared wavelengths. Two candidate lasers are lead salt diode lasers and modulation sideband CO_2 and CO lasers. Another widely used family of tunable infrared lasers are the optical parametric oscillators. However, these generally operate only in a pulsed mode, and we prefer a cw source for our experiment as cw sources can be made more monochromatic and they permit the measurement of the entire decay profile following each photolysis pulse.

We have used cryogenically-cooled lead salt diode laser probes. However, we stopped using them a few years ago because of various problems that we encountered. They may work much better now, but, at the time we decided to quit using them, we were having a lot of trouble. Line tunable CO and CO_2 lasers can be tuned over small frequency regions by adding side bands. These can be highly satisfactory probe lasers, but they are not conveniently broadly tunable.

Difference Frequency Generation

Other tunable cw infrared sources are based on difference frequency generation. (*3*) We became interested in such sources as a replacement for lead salt diodes. In difference frequency generation, a high frequency laser, "the pump," and an intermediate frequency laser, "the signal," are mixed in a nonlinear medium producing

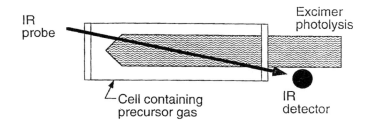

Figure 1.   The infrared kinetic spectroscopy experiment. In the experiment a suitable precursor is flash photolyzed by an excimer laser (typically ArF at 193 nm) to produce free radicals. The transient infrared absorption of the radical is observed by acquiring the infrared detector output with a transient digitizer.

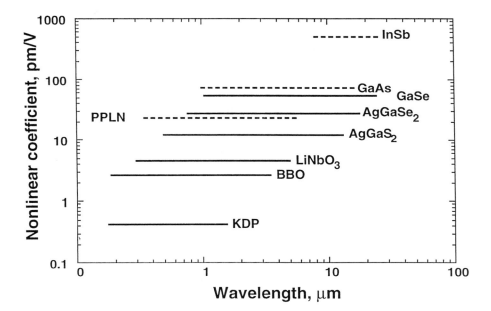

Figure 2.   Nonlinear mixing materials. The locations of the horizontal bars show the wavelength region where the material is transparent. The vertical axis is the nonlinear mixing coefficient. The infrared power obtainable for a given level of pump and signal power is very approximately proportional to the square of this coefficient. The horizontal bars that appear as broken lines cannot be phasematched by crystal birefringence, but in principle all can be quasiphasematched and in the case of LiNb03 indicated by PPLN quasiphasematching has been demonstrated to be practical and is used routinely.

infrared light ("the idler") at a frequency equal to the difference between the pump and signal frequencies.

Consider what happens as the laser beams travel down the crystal. After some distance, there will be an idler electric field already present as a result of non-linear mixing in the region of the crystal already traversed. Over a short distance following this point, an increment to the idler electric field is generated by the mixing of the electric fields associated with the pump and signal. If the sign of this additional field is the same as the electric field already present, the idler wave continues to build up, but if the sign of incremental field is opposite to that already present the idler wave actually decreases. One wants the idler to build up; this requirement that the incremental field always add imposes a condition on the refractive indices at the three wavelengths called "the phase matching condition." Normally, the non-linear materials used are birefringent having an index of refraction that depends upon the light polarization, and this birefringence can make it possible to satisfy the phase matching condition. However, for a given idler frequency, the phase matching requirement, if it can be satisfied at all, often fixes the absolute frequencies of the pump and signal reducing the flexibility of choice of these lasers.

There are several choices of non-linear mixing crystals. Figure 2 shows some of the relevant characteristics of the materials which we believe are most suitable. Three properties of a material determine its practical utility as a mixing material to generate infrared light. First, it must be transparent at the desired infrared wavelength *and* at the pump and signal wavelengths. Therefore, in Figure 2, the horizontal axis is wavelength and the bars are showing the wavelength region over which the material is transparent. Second, the non-linear mixing process must be reasonably efficient. The vertical axis in the figure is the non-linear coefficient. As a rough rule of thumb, the amount of infrared power produced for a given amount of pump and signal power is proportional to the square of the non-linear coefficient. Because of this squaring, big increases in infrared output power can be expected in moving from materials at the bottom of the figure to materials near the top for the same pump and signal powers. Third, it must be possible to satisfy the phase matching condition. The materials with the dotted bars are materials where it is not possible to satisfy the phase matching condition through the normal utilization of birefringence. Nevertheless as we shall see, they can be useful.

We have used silver gallium sulfide ($AgGaS_2$), silver gallium selenide ($AgGaSe_2$), and gallium selenide (GaSe) to generate single frequency cw infrared We have worked most recently with the dotted line material periodically-polled lithium niobate (PPLN). For $AgGaS_2$ and $AgGaSe_2$ because of the phase matching condition, optimum performance requires a certain value for the pump and signal frequency for a given infrared frequency. In the case of $AgGaS_2$, the pump and signal are conveniently located in the region from 600-900 nm where cw dye and Ti:sapphire lasers have been highly developed. We have found that difference frequency mixing in $AgGaS_2$ provides a quite satisfactory laboratory infrared source. (*4, 5*)

However, for $AgGaSe_2$, which is otherwise a much more interesting material because it has a higher nonlinear coefficient and is transparent to longer wavelengths in the infrared, the pump and signal must be in the 1000-1600 nm region. There are fewer cw sources in this region and none commercially available that are broadly tunable. Thus our experiments with this material were much more limited in scope. (*6*)

The crystal structure of GaSe is such that it can never be optimally phase matched, but the best phasematching is obtained with convenient Ti:sapphire sources for the pump and signal and IR powers of practical interest can be generated. (7)

PPLN (8, 9) is an example of a quasiphasematched material. In quasiphasematching, the direction of the optic axis is reversed in striations in a direction perpendicular to the direction of the laser beams in such a way that the sign of the infrared electric field being generated reverses when the laser beams pass from one striation to the next. By making the width of the striations such that the sign reversal takes place just when the electric field of the infrared being generated begins to subtract rather than add, the infrared beam is made to build up continuously going through the material. The net result is that, if the material is amenable to this approach, one can put the phasematching into the material fabrication and the only restriction on the pump and signal frequencies is that their difference must be the desired infrared frequency. This permits the use of any convenient laser combination. Even a fixed frequency laser can be used for the pump or the signal by creating crystals with a rows of channels across the crystal. The channels are oriented in the propagation direction with each channel having a slightly different striation width.

DFG-based Portable Infrared Monitors

After we started working with these materials, my colleague, Frank Tittel, who is a laser engineer, had a brilliant idea. He recognized that there are many visible and near infrared diodes that are highly developed. Everybody has one in their CD player or their laser pointer. These diode lasers also are used for optical fiber communication. They are cheap and they are efficient; up to twenty-five percent of the electrical energy that comes out of the wall becomes light energy out of the diode laser. In addition there are some powerful and reasonably efficient fixed frequency lasers such the Nd:YAG that can be used very effectively in the quasiphasematched scheme. Figure 3 shows the power ratings and wavelength regions of the diode and fixed frequency lasers. In essence, Frank said, "These lasers are small, they are efficient, they are light. We can build small portable devices based on mixing their outputs to generate infrared for the purposes of environmental monitoring. These devices will operate without cryogenic cooling or large power requirements"

We have started in this direction. Figure 4 shows a schematic of a monitor configured for open-path monitoring. The system starts with two relatively efficient diode lasers. The mixing in the non-linear crystal is highly inefficient so that not much infrared power is produced, but you get enough out to be able to do spectroscopy quite conveniently. Single frequency tunable infrared light is produced meaning that no matter how narrow the infrared absorption line to be observed is, it is still broader than the laser bandwidth.

Monitoring can be done two ways: by sampling into a cell or in an open path. Open path monitoring can be very useful. For example, if you wish to measure the carbon monoxide emission of automobiles, you can put a monitor across the freeway ramps, and measure how much carbon monoxide the cars are emitting as they go by thereby providing a means for policing emissions. The open path can be along the fence line of the chemical plant. Chemical plants are very interested in doing this in order to make sure that dangerous emissions are not escaping from the plant. This

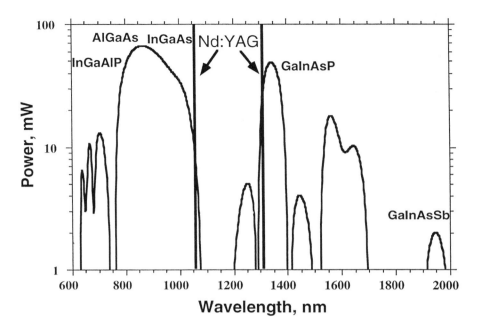

Figure 3.   Currently available diode lasers showing the frequency region covered and the available power. The two lines of the Nd:YAG laser are shown as vertical lines.

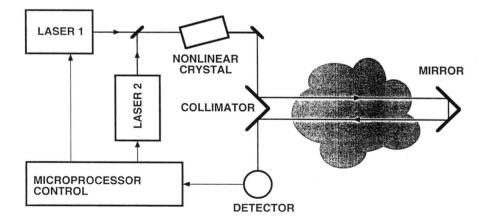

Figure 4.   Schematic of monitor based upon difference frequency generation in the open path configuration.

kind of monitoring is also very useful for legal purposes. If the people in the subdivision across the road from the plant sue them some years later claiming that they have got cancer because of something that came out of the plant, chemical manufacturers would like to be able so say, "No, we've been monitoring our fenceline for years and years and none of this stuff has come across the road." Or, such a device could be used to monitor the emissions from a volcano. Volcanologists are very interested in what is is being emitted from the fissures in a volcano in the hope that these emissions can be correlated with eruptions with the ultimate aim of predicting eruptions.

As an example of the open path monitor in operation, Figure 5 shows an absorption line of carbon monoxide as seen in ambient concentration in air in a pathlength of four meters. (10) The concentration of carbon monoxide causing this signal corresponds to about three hundred parts per billion in air. Many species can be observed at ambient levels with such monitors; however, lines observed in open path at atmospheric pressure are relatively broad because of pressure broadening and in many cases lines of different species found in the atmosphere overlap. Narrower lines with reduced possibility of overlapping lines can be obtained using point sampling.

Figure 6 shows the same apparatus modified for point sampling. It uses the same infrared source, but instead of the absorption path running through air, the infrared absorption takes place in a small multi-path cell. At pressures above about 100 Torr, the width of an absorption line is determined by pressure broadening and is proportional to pressure. The integrated intensity of a line is proportional to the amount of substance present. The same peak intensity of the infrared absorption line with a much narrower line width is obtained just by pumping the air out of the cell until the pressure is reduced to about 100 Torr, because the reduction in amount of absorber is compensated by the narrowing of the line. Continued pumping to pressures below about 100 Torr reduces the peak absorption because the linewidth is determined by Doppler broadening rather than pressure broadening. As shown in Fig. 7, when the pressure is reduced the absorption lines become narrower, but the peak intensity remains the same.

The engineering required to make a portable gas monitor based upon difference frequency is fairly well developed. Figure 8 shows a photograph of a portable infrared monitor constructed by Konstantin Petrov, Thomas Töpfer, and David Lancaster in Frank Tittel's lab for monitoring carbon monoxide, methane and formaldehyde. (11) This monitor is clearly portable defining portable as something that a strong person can carry for miles. This monitor includes all its power supplies, but not the actual source of power. However, it is not rugged. If it is carried for miles and then put into service, it will probably have to be realigned optically before it will operate. Also the optics may require cleaning, even in a low dust environment. By replacing the optic train for the pump and signal with fiber optics, the need for realignment and cleaning can be eliminated. This will be the next step taken to make this monitor more rugged.

The step after that will be the use of waveguide PPLN, an advance pioneered by Marty Fejer's group at Stanford. (12) By producing an optical waveguide in the material, the three frequencies can be kept confined in a very small cross-sectional area down the entire length of the crystal. This confinement provides a much higher conversion efficiency, so that much less powerful lasers can be used for the difference frequency generation to produce the same infrared power. Consequently, the

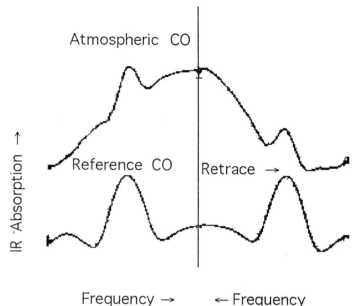

Figure 5. Wavelength-modulation $2f$ spectra of the R(6) fundamental of CO at 2169 cm^{-1} in 4.2 m ambient air (x30 magnified, top), and a reference sample (~3 torr CO mixed with room air in a 10 cm cell, bottom). Frequency sweep of ~20 GHz is reversed at the center. Sweep rate was 10.6 Hz, modulation frequency was 2kHz, lock-in time constant was 1 ms. Both traces are 100 sweep averages.

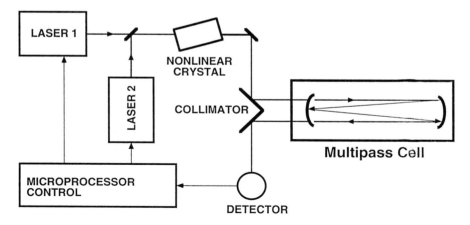

Figure 6. Schematic of monitor based upon difference frequency generation in the point sampling configuration.

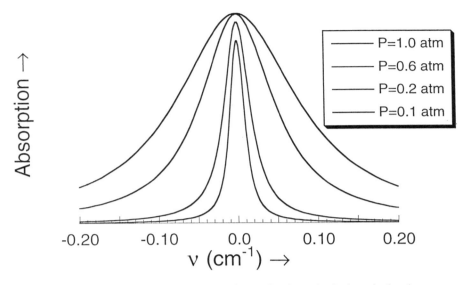

Figure 7.   Calculated absorption line profiles of a hypothetical typical substance at several reduced pressures compared. As the cell is pumped out reducing the pressure, the peak intensity remains almost constant until the linewidth begins to be determined by Doppler broadening rather than pressure broadening around 100 Torr. The loss of sample concentration as the result of loss of material is compensated by the line narrowing so that the peak intensity remains nearly constant.

Figure 8.   Photograph of a portable infrared monitor based upon difference frequency generation.

instrument can be constructed with less expensive lasers requiring less power which will be cheaper and lighter. An instrument that can be carried any place and be expected to work with minimal adjustments for monitoring a specific gas can be developed with a production line price of a few thousand dollars.

Ground NASA Missions

So, what's the big deal. What I have described so far is just a simple exercise in analytical chemistry. It is maybe a little higher tech than some analytical instruments depending as it does on technology that is still in development, but that's all. What's the point? Let me tell you about what this is being used for now as a parable for more global concerns. NASA is very interested in long duration space missions. The immediate need is for the Space Station, and very long term they are still interested in moon bases and going to Mars, even if Congress is not. In order to develop advanced space cabin environments and anticipate problems of human life support during a long mission, they are running mock missions in which four people are locked in a chamber at Johnson Space Craft Center.

In one of the previous experiments, the test almost had to be aborted because the levels of formaldehyde became too high in the chamber. Concentrations reached over a part per million; the eyes of the test subjects became irritated. Thus there is a clear need to eliminate formaldehyde. We have been helping them by studying samples of different plastics that go into these chambers to see if they emit formaldehyde using our infrared monitor (13) and have found that a lot of materials that they were contemplating putting into the chamber emit copious amounts of formaldehyde. The chamber design has been modified to eliminate these materials.

In September 1997, the four people were locked up into this chamber. They were released in mid-December after serving three months. These individuals went in happily, and seemed to enjoy this mock mission even though as far as I am concerned it would be like going to jail. They were excited to be involved in this project. The serious purpose is to simulate the closed environment of a spaceship on a long-term mission in order to ascertain what problems might arise. In an effort to produce a microecology, the earth-bound astronauts actually grew wheat in this chamber. David Lancaster and Dirk Richter used our monitor to follow the formaldehyde concentration in the chamber. Fortunately the use of non-formaldehyde emitting materials in the chamber construction and the introduction of a chemical scrubber kept the formaldehyde concentrations at levels far below those causing irritation.

Spaceship Earth has a Problem

These NASA space chamber experiments remind us that we are on a spaceship called "earth," and we have a problem. The problem is the growth of world population. About twelve thousand years ago, there might have been about six million people on this planet. (14) At the beginning of the present era, the world population was about two hundred fifty million people. (14) The growth from six million to over than 250 million over 10000 years can be associated with the development of agriculture, which began roughly ten thousand years ago. In the present era, world population grew slowly until about 1750 when it reached nearly 800 million. Then the Industrial Revolution and the introduction of science and technology provided the resources that increased life expectancy at birth from about twenty seven years in 1750 to about 35 years in 1950. (*14*) Instead of dying of bubonic plague, smallpox, and childbirth fever, people lived through their reproductive years, and reproduced. The result is that the population has taken off like a rocket rising to 2.5 billion in 1950 and nearly 6 billion today. Figure 9 shows these long term population trends.

Our spaceship is in trouble. Figure 10 shows the population projections to roughly 2025. (*15*) This upward curve has to stop somewhere because the earth will not be able to support any more people. The question is how will it stop? Is it going to shoot up, reach a maximum, and then plummet indicating some catastrophe? About the only part of the future that can really be predicted is that the world population is going to increase unless there is a catastrophe. Where population growth will stop, nobody can tell. The best estimators think it is going to level off around 2050 at a population of about ten billion people. I do not know whether that estimate is rational, or contains an element of wishful thinking. It has to level off because the earth can support only a finite population, but the major issue is what that level will be.

Most of the growth in recent years is coming from the underdeveloped countries. The developed country populations are stabilizing and leveling off; any growth is rather slow. Underdeveloped countries are experiencing almost runaway growth. This disparity between population growth in the developed and underdeveloped countries means that the world is going to change. Most people are going to live in underdeveloped countries by a large majority, maybe seven to one, in the year 2025.

We are all in this together. We cannot as a nation, say, "This is not my problem," because air and sea pollution through the emission of various noxious gases into the air and chemicals into the sea in the rest of the world will adversely affect us. The world will inevitably be changed globally by mankind. If we keep our own house in order, it is going to help us. Los Angeles smog is not as bad today as it was twenty or twenty-five years ago. However, if we have made some progress in some areas, we are not making progress everywhere even in the U.S. I was just recently in Columbus, Ohio. The smog was terrible. Whatever we do at home, however, is not going to prevent us being affected by the rest of the world. There is terrible smog in most places that I go in the world e.g. the Far East, Europe. The only place that I have been recently, where there wasn't any smog problem, was the South Island of New Zealand. They are sort of away from it all and the population density is low. Except the Kiwis are worried about the ozone hole growing to expose them. You cannot get away from man's global effect on the world even on the South Island of New Zealand.

Global Warming

Consider the global warming problem. Figure 11 shows the global temperature variation according to the Vostok glacier isotopic temperature record. (*16*) Everybody knows that mankind lived through an Ice Age about 20000 years ago. In addition humanity lived through a hot spot time about 130,000 years ago when the global temperatures were somewhat higher than at present. There are not any reports on how life was when the world was warmer. Currently, we are experiencing another warm

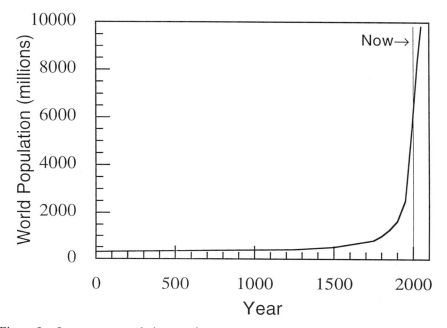

Figure 9.   Long term population trends.

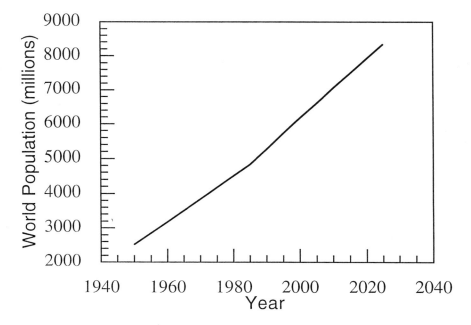

Figure 10. Population 1950 to 2025.

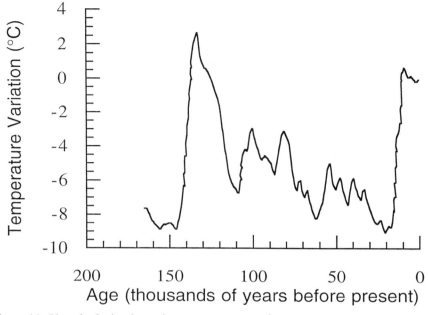

Figure 11. Vostok glacier isotopic temperature record.

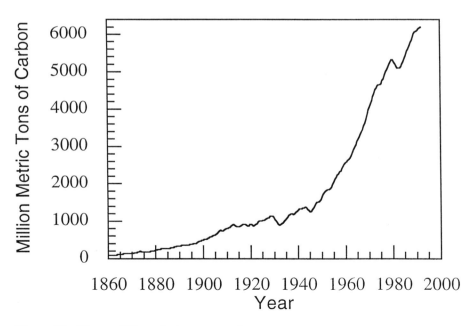

Figure 12. Human CO_2 emissions versus time.

era. Now on top of the warm era caused by natural climate changes, we are adding a warming force through the emission of greenhouse gases.

Most people focus on CO_2. Figure 12 shows a plot of world CO_2 emissions in terms of metric tons of carbon emitted versus time. (17) It is clear that mankind is putting out increasing amounts of carbon in the form of carbon dioxide. Is it really building up? The answer is, yes. Figure 13 is graph of the carbon dioxide concentration at Mauna Loa, which should reflect world levels fairly well, versus time. (18) It is clearly rising, and the curve appears to be concave upwards suggesting that the problem is getting worse more rapidly.

The increased concentrations of gases that absorb in the infrared absorb the infrared light radiated from the earth's surface. These greenhouse molecules then reradiate the energy isotropically in the infrared. The light radiated upward is again absorbed and reradiated until eventually the heat is radiated into space. The net effect is that the atmosphere acts like thermal insulation keeping the heat in. Initially the stratosphere will cool because it is receiving less heat from below, and the surface temperature will rise. At the new steady state heat balance, the stratospheric temperatures will rise to their old levels and the average surface temperatures will be higher. The steady state global temperature increase based upon this simple picture is fairly easily calculable amounting to a temperature increase of about 1.2 °C upon doubling of our present CO_2 concentrations. (19) However, uncertainties arise because the greenhouse climate forcing function comes with feedbacks, both positive feedback enhancing the warming effect and negative feedback decreasing the global warming effect of the greenhouse gases. Because of these feedback effects, the estimation of the expected temperature rise depends upon complex climate models and the estimated temperature rise for doubled CO_2 ranges from 2 to 5 °C depending upon the model.(19) The fact that the 2 °C lower number is higher than the 1.2°C calculated without feedbacks indicates that all climate modelers believe that positive feedback is dominant.

One can look at recent history. It is estimated(20) that global CO_2 concentrations have risen about 40% since the 1880's. This would project to a temperature rise of 1 to 2 °C at equilibrium. In fact, the best estimate(20) is that mean land temperatures have risen about 0.5 °C, and there is some uncertainty about whether the rise is real and can be attributed to CO_2. However, this observation doesn't get us off the hook. The global system responds very slowly to forces driving climate change taking decades even perhaps a century to approach equilibrium.(21) The need for research aimed at understanding the causes of global temperature variation and the importance of greenhouse gases in causing such variations remains urgent.

At present, we are caught in frightful dilemma. Without being quite sure how severe the effects of our emissions of greenhouse gases are, we must act and even make sacrifices because humanity cannot afford the consequences if the worst predictions were to come true. The long time constant of climate change means that we cannot expect that warming will stop as soon as we stop emitting greenhouse gases.

Carbon dioxide is not the only greenhouse gas. Methane is of concern because its infrared absorptions are at different wavelengths from the carbon dioxide absorptions so that it fills in some of the holes where the air was transparent. Figure 14

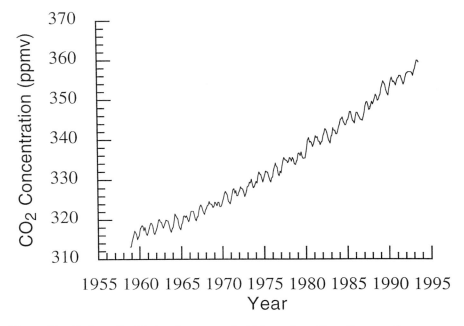

Figure 13.   Carbon dioxide levels measured at Mauna Loa, Hawaii versus time.

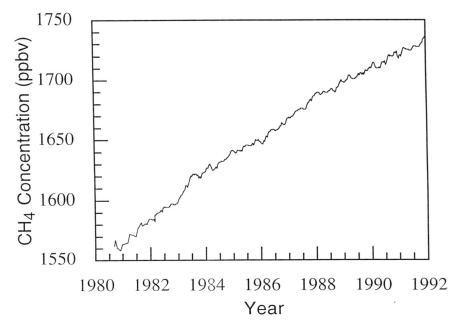

Figure 14.   Methane global concentrations versus time.

depicts methane concentration versus time.(22) Obviously, the concentration of methane is rising. In contrast with Figure 11, Fig. 14 is somewhat concave downward; maybe we should be somewhat encouraged, maybe not.

Methane from Rice Farming and IR Monitors

Where is the excess methane originating? Table 1 shows the sources of methane.(23) The total emissions, natural and man-caused, are five hundred teragrams per year and the loss of methane per year is four hundred and sixty teragrams per year. This means that a net of forty teragrams of methane is being added every year to the global atmosphere. Mankind produces more than twice the methane arising from natural sources. The largest man-made source is non-agricultural, primarily from coal, petroleum, and natural gas production. The next largest source is cattle raising, and the next largest source results from rice cultivation. Why focus on rice fields rather than cattle raising? When there are ten billion people on the planet, there are going to be few who can afford the luxury of beef. In order to feed more mouths, the production of rice is undoubtedly going to increase. The raising of cattle probably will not increase much.

For years, Ronald Sass has been setting up small air traps at various points over the rice fields and taking the collected air samples to the lab where they are analyzed for methane. As a biologist, he wants to understand what cultivation conditions and organisms (The organisms producing methane in rice cultivation are generally anerobic soil bacteria) are producing methane. We are working with him to build a detector that is going to be rugged and portable. We hope rugged enough that, if you slip and drop it in the rice paddy, it will still work. Such a detector will help a lot to make the work of methane flux measurement go faster because by using open path monitoring, you can cover a much wider area.

Table 1 Methane Sources & Sinks

Natural	160 Tg/yr
Anthropogenic	340
– non-agriculture	100
– Enteric (cattle)	80
– Flooded ricefields	50
– Other	110
Total Source	500
Sinks	-460
NET	40 Tg/yr

R. L. Sass & F. L. Fisher, 1996 (ref. 23)

Summary

Let's return to the big picture. The world population is definitely going to grow and probably will nearly double within the next thirty years. The additional people are going to need energy, they are going to need food, and humanity must not drown, or cook, in its own waste. This implies that three important areas for research for the future will be energy production, agriculture and environmental work.

A few years ago in this country, pressures were building to justify research in terms of improving economic competitiveness. From that point of view, the supporters of basic research had no convincing reply to an argument that "If we do basic research, it will get published and will go all over the world. Then other countries apply our basic research and basic research will not have helped our economic competitiveness at all."

Over the last few years, the world has changed. Instead of thinking of research just in terms of economic competitiveness, we now are beginning to recognize that all humanity has common problems that will require technological solutions. To develop the knowledge base needed to find these solutions, a tremendous amount of both basic and applied research is required. As I have indicated above, humanity's problems will inevitably become the USA's problems. We must work on efficient new methods of air pollution abatement merely to avoid being choked by the fumes produced by billions of additional people. We must find energy sources that emit less carbon dioxide to avoid global warming. We must continue and increase agricultural research to avoid a world containing billions of starving or severely undernourished people. Inevitably, this country will be forced to support basic research for the good of mankind. We are going to have to do it so that we can have, or our children can have, a life that's worth living.

Literature Cited

1. Adamson, J. D.; DeSain, J. D.; Curl, R. F.; Glass, G. P. *J. Phys. Chem.* **1997**, *101*, p 864.

2. Stephens, J. W.; Morter, C. L.; Farhat, S. K.; Glass, G. P.; Curl, R. F. *J. Phys. Chem.* **1993**, *97*, p 8944.

3. Pine, A. S. *J. Opt. Soc. Am.* **1974**, *64*, p 1683.

4. Canarelli, P.; Benko, Z.; Curl, R. F.; Tittel, F. K. *J. Opt. Soc. Am. B* **1992**, *9*, p 197.

5. Hielscher, A. H.; Miller, C. E.; Bayard, D. C.; Simon, U.; Smolka, K. P.; Curl, R. F.; Tittel, F. K. *J. Opt. Soc. B* **1992**, *9*, p 1962.

6. Petrov, K. P.; Curl, R. F.; Tittel, F. K.; Goldberg, L. *Optics Lett.* **1996**, *100*, p 8008.

7. Eckhoff, W. C.; Putnam, R. S.; Wang, S.; Curl, R. F.; Tittel, F. K. *Appl. Phys. B* **1996**, *63*, p 437.

8. Sanders, S.; Lang, R. J.; Myers, L. E.; Fejer, M. M.; Byer, R. L. *Proceedings CLEO'95* **1995**, p 370.

9. Goldberg, L.; Burns, W. K.; McElhanon, R. W. *Appl. Phys. Lett.* **1995**, *67*, p 2910.

10. Petrov, K. P.; Goldberg, L.; Burns, W. K.; Curl, R. F.; Tittel, F. K. *Opt. Lett.* **1996**, *21*, p 86.

11. Töpfer, T.; Petrov, K. P.; Mine, Y.; Jundt, D.; Curl, R. F.; Tittel, F. K. *Appl. Optics* **1997**, *36*, p 8042.

12. Arbore, M. A.; Chou, M.-H.; Fejer, M. M. *CLEO '96 Proceedings (Opt. Soc. Am., 1996)* **1996**, p 120.

13. Mine, Y.; Melander, N.; Richter, D.; Lancaster, D. G.; Petrov, K. P.; Curl, R. F.; Tittel, F. K. *Appl. Phys. B* **1997**, *65*, p 771.

14. Livi-Bacci, M. *A Concise History of World Population*; Blackwell: Oxford, 1992.

15. Department for Economic and Social Information and Policy Analysis, Population Division, United Nations, *World Population Prospects: The 1994 Revision*; United Nations: New York, 1995.

16. Jouzel, J.; Lorius, C.; Petit, J. R.; Barkov, N. I.; Kotlyakov, V. M. *Vostok Isotopic Temperature Record*; Boden, T. A.; Kaiser, D. P.; Sepanski, R. J.; Stoss, F. W., Ed.; Oak Ridge National Laboratory: Oak Ridge, TN, 1994.

17. *Trends '93: A Compendium of Data on Global Change*; Boden, T. A.; Kaiser, D. P.; Sepanski, R. J.; Stoss, F. W., Ed.; Oak Ridge National Laboratory: Oak Ridge, TN, 1994.

18. Keeling, C. D.; Whorf, T. P. Atmospheric CO_2 Records from Sites in the SIO Air Sampling Network,*Trends '93: A Compendium of Data on Global Change*; Boden, T. A.; Kaiser, D. P.; Sepanski, R. J.; Stoss, F. W., Ed.; Oak Ridge National Laboratory: Oak Ridge, TN, 1994.

19. Hansen, J.; Lacis, A.; Ruedy, R.; Sato, M.; Wilson, H. *National Geographic Research and Exploration* **1993**, *9*, p 143.

20. Barnett, T. P. *et al Working Group III: Greenhouse Signal Detection*; Barnett, T. P., Ed.; Elsevier: Amsterdam, 1991; Vol. 19, p 594.

21. Hansen, J. E.; Lacis, A. A. *Nature* **1990**, *346*, p 713.

22. Khalil, M. A. K.; Rasmussen, R. A. Global CH_4 Record Derived from Six Globally Distributed Locations, *Trends '93: A Compendium of Data on Global Change*; Boden, T. A.; Kaiser, D. P.; Sepanski, R. J.; Stoss, F. W., Ed.; Oak Ridge National Laboratory: Oak Ridge, TN, 1994.

23. Sass, R. L.; Fisher, F. M. *Current Topics In Wetland Biogeochemistry* **1996**, *2*, p 24.

4

Oil and Hydrocarbons in the 21st Century

George A. Olah

Director, Loker Hydrocarbon Research Institute
University of Southern California, Los Angeles, CA 90089-1661

The population of the earth as we are nearing the 21st Century is six billion. Even if mankind would increasingly exercise some form of population control, in a quarter of a century we probably will reach around 9.5-10 billion. This inevitably will put enormous pressure on our resources, not the least our energy resources. When hydrocarbons, including fossil fuels, are burned, they produce carbon dioxide, CO_2, and water, H_2O. This is the reason why hydrocarbon resources are non-renewable. A challenging new approach is to reverse the process and produce hydrocarbons from carbon dioxide and water. In the laboratory, we already know how to convert carbon dioxide back into hydrocarbons through chemistry using hydrogen gas, H_2. Our superacid chemistry made significant progress to bring about the feasibility of such approach. The limiting step at this time is the electricity needed for generating hydrogen from water. Atomic power plants, *albeit* improved and made safer, will eventually give us needed cheap energy. At the same time, we still can not store electricity efficiently and thus, fossil fuel burning power plants in off peak periods could produce hydrogen in a way storing energy. Other approaches to cleave water may also be utilized, such as enzymatic. In cooperation with JPL, we developed, in recent years, a new direct oxidation liquid feed fuel cell using methanol. The cell produces CO_2, H_2O and electricity. We are now also working on the reverse process, i.e. producing methyl alcohol from CO_2 and H_2O in an electrocatalytic way. As CO_2 is also a significant green house gas responsible for global warming, the side benefit of the technology of recycling CO_2 into a useful general fuel is to mitigate this serious environmental problem.

World Energy Consumption into the 21st Century

The rapidly growing world population which was 1.6 billion at the beginning of the 20th century has now reached 6 billion.

World Population (in millions)							
1650	1750	1800	1850	1900	1920	1952	2000
545	728	906	1,171	1,608	1,813	2,409	6,200

With an increasingly technological society the world's per capita resources have difficulty in keeping up. Society's demands need to be satisfied while safeguarding the environment and allowing future generations to continue to enjoy planet earth as a hospitable home. To establish an equilibrium between providing for mankind's needs while safeguarding and improving the environment is one of the major challenges of society. Men need not only food, water, shelter, clothing and many other perquisites, but also energy. (Fig. 1)

Our early ancestors discovered fire and started to burn wood. The industrial revolution was fueled by coal and the 20th century added oil and gas, and introduced atomic energy.

A Real Energy Crisis by the Second Half of the 21st Century

When fossil fuels such as coal, oil or natural gas (i.e. hydrocarbons) are burned as fuels in power plants to generate electricity or to heat our houses, fuel our cars, airplanes, etc., they form carbon dioxide and water. Thus, they are used up and are non-renewable.

Fossil Fuels

Petroleum Oil, Natural Gas, Tar-Sand, Shale Bitumen, Oil, Coals

They are mixtures of hydrocarbons, i.e. compounds of the elements carbon and hydrogen. When oxidized (combusted) they form carbon dioxide (CO_2) and water (H_2O) and thus are not renewable (on the human time-scale).

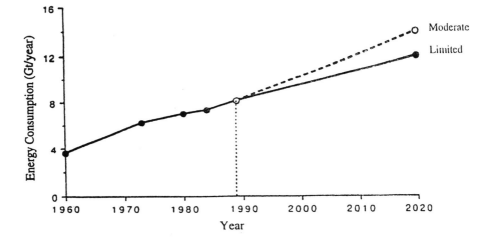

Figure 1. World energy consumption (in gigatons/year) projections.

Nature has given us a remarkable gift in the form of oil and natural gas. What was created, however, over the ages, man is using up rather rapidly. The large scale use of petroleum and natural gas to generate energy, and also for raw materials for diverse man-made materials and products (fuels, plastics, pharmaceuticals, dyes, etc.) all developed during the 20th Century.(*1*) The US energy consumption is very heavily based on fossil fuels. Atomic energy and other sources (hydro, geothermal, solar) represent only 11-12%.

	U.S. Energy Sources (%)		
Power Source	1960	1970	1990
Oil	48	46	41
Natural Gas	26	26	24
Coal	19	19	23
Nuclear Energy	3	5	8
Hydro-, Geothermal, Solar, etc. energy	4	4	4

Other industrial countries in contrast get 25 to 85% of their energy from non-fossil sources.

Power Generated in Industrial Countries by Nonfossil Fuels (1990)			
	Nonfossil Fuel Power %		
Country	Hydroenergy	Nuclear Energy	Total
France	12	75	87
Canada	58	16	74
Former West Germany	4	34	38
Japan	11	26	37
UK	1	23	24
Italy	16	0	16
USA	4	8	12

Oil use has grown to the point where the world consumption is approaching 55-60 million barrels (a barrel equals 42 gallons, i.e. some 160 liters) a day or some 10 million metric tons. Oil and gas are mixtures of hydrocarbons. As already mentioned, once we burn hydrocarbons, they are irreversibly used up and are not renewable on the human time scale. Fortunately, we still have significant worldwide reserves, including heavy oils, shale and tar-sands and even larger deposits of coals (a complex mixture of carbon compounds more deficient in hydrogen) which can be eventually utilized, *albeit* at a higher cost. I am not suggesting that our resources will run out in the foreseeable future, but it is clear that they will become scarcer, much more expensive, and won't last for very long. With the world population at 6 billion and rapidly growing (it may reach 10 billion in a few decades), the demand for oil and gas can only increase. It is true that, in the past, dire predictions of fast disappearing oil and gas reserves were always incorrect.

Recognized Oil and Natural Gas Reserves (in billion tons) from 1960 to 1990		
Year	Oil	Natural Gas
1960	43.0	15.3
1965	50.0	22.4
1970	77.7	33.3
1975	87.4	55.0
1980	90.6	69.8
1986	95.2	86.9
1987	121.2	91.4
1988	123.8	95.2
1989	136.8	96.2
1990	136.5	107.5

The question is, however, what is "fast" and what is the real extent of our reserves. Proven oil reserves instead of being depleted, as a matter of fact, nearly doubled in the last 30 years and now exceed a trillion barrels. This seems so impressive that most people assume that there is no oil shortage in sight. However, the increasing consumption coupled with a growing world population makes it more realistic to consider per capita reserves. If we do this, it becomes evident that our known reserves can last for no more than half a century. Even if we consider all other factors (new findings, savings, alternate sources, etc.) in the first half of the 21st Century, we will increasingly face a major problem. Oil and gas will not become exhausted overnight but market forces of supply and demand will inevitably start to drive prices up to levels no one even wants to contemplate presently. Even so, by the second half of the century, if we don't find new solutions, we will face a real crisis.(1)

All mankind wants the advantages an industrialized society can give its citizens. We essentially rely on energy, but the level of consumption is vastly different in different parts of the world (industrialized vs. developing world). Oil consumption per capita, for example, in China presently is only 5 barrels/year, but it is more than ten fold in the US. China's oil use is expected under the most conservative estimates to double in the next decade. This alone equals the US consumption, reminding us of the size of the problem we are facing. Imagine the further demand if the Chinese (and others) will not be satisfied riding bicycles but would increasingly expect to drive cars and use other conveniences common in the developed countries. Do we in the industrialized world have a monopoly for better life? I certainly don't think so.

Approaches for New Energy Sources

Atomic Energy. Generating energy by burning non-renewable fossil fuels including oil, gas and coal is feasible only for the relatively short future and even so, faces serious environmental problems (*vide infra*). The advent of the atomic age opened up a wonderful new possibility, but also created dangers and concerns of safety. I feel that it is tragic that the latter considerations practically brought further development of atomic energy to a stand still at least in most of the Western world. Whether we like it or not we have in the long run no alternative but to rely increasingly on clean atomic energy, but we must solve safety problems including those of disposal and storage of radioactive waste-products. Pointing out difficulties and hazards as well as regulating them (within reason) is necessary. Finding solutions to overcome them, however, is essential.

Synthetic Oil Products. If we continue to burn our hydrocarbon reserves to generate energy and to use them as fuels, etc., diminishing resources and sharply increasing prices in the 21st Century will lead inevitably to the need to supplement or make them ourselves by synthetic manufacturing. Synthetic gasoline or oil products will be, however, much costlier. Nature's petroleum oil and natural gas are the greatest bargains we will ever have. A barrel of oil still costs only around $20 (with some market fluctuation). No synthetic manufacturing process will be able to come even close to this price and we will need to get used to this, not as a matter of any government policy, but as a fact of market forces over which we have little control.

Synthetic oil is feasible and can be produced from coal or natural gas via synthesis-gas (a mixture of carbon monoxide and hydrogen obtained from incomplete combustion of coal or natural gas which are, however, as pointed out, non-renewable resources). Coal conversion was used in Germany during World War II and in South Africa during the boycott years. However, the size of these operations hardly amounted to 0.3% of the present US consumption. This route (the so called Fischer-Tropsch synthesis) which is also highly energy consuming, gives unsatisfactory product mixtures and hardly can be the technology of the future. New and more economical processes are needed. Some of the needed basic science and technology is evolving. For example, more abundant natural gas can be directly converted without first producing the synthesis-gas to gasoline or hydrocarbon products. Using coal and natural gas to produce oil will extend its availability, but new approaches based on renewable resources are essential for the future.

Recycling Carbon Dioxide to Produce Energy

As mentioned, when hydrocarbons are burned they produce carbon dioxide, CO_2, and water, H_2O. It is a challenge for the future to reverse this process and to produce efficiently and economically hydrocarbon fuels from carbon dioxide and water. This may sound like science fiction, but it is not. In principle, we chemists already know how to convert carbon dioxide with hydrogen gas, H_2, into methyl alcohol. Catalytic processes using metal or highly acidic catalysts can be used for this conversion. The limiting factor is that to produce needed hydrogen (such as by electrolytically splitting water) much energy is needed. In the long range, this can be provided by atomic energy, *albeit* improved and made safer. Alternative energy sources can also contribute. Use of photovoltaic solar energy is possible in suitable locations, such as desert areas. Energy of the wind, waves, tides, etc. can potentially also be used, but all these sources are far from economical or feasible. At present, our existing power plants, either burning fossil fuels (i.e. oil, gas and coal) or using atomic energy, have substantial excess capacity in off peak periods as we still can not store electricity efficiently. Our existing power plants in off peak periods even now could produce hydrogen, in a way storing energy. Other means to cleave water may also evolve such as the use of enzymes or using sunshine for energy. Nature, of course, recycles CO_2 by photosynthesis in plants, trees, algaes of the oceans, etc. to carbohydrates and cellulose, thus renewing plant life. (Fig. 2)

Some plants even produce hydrocarbons, such as natural rubber, but the scale needed for a significant contribution is enormous. What I am foreseeing is supplementing nature and producing synthetic hydrocarbons from carbon dioxide and water through chemistry on a large scale in new efficient and economical ways.

A practical approach would involve recycling carbon dioxide from industrial emissions to usable new hydrocarbon fuels. The average carbon dioxide content of the atmosphere is very low (0.035%) and, therefore, CO_2 is difficult to separate from the air economically. The noble gas argon is present in nearly 30 fold amount in our air, but CO_2 plays an essential role in our terrestrial life.

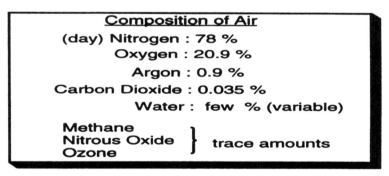

However, CO_2 can be readily recovered from emissions of power plants burning carbonaceous fuels (coal, oil, natural gas), from fermentation processes, and from calcination of limestone or other industrial sources.

Carbon Dioxide and the Green House Warming Effect

As power plants and other industries emit large amounts of carbon dioxide, they contribute to the so-called green house warming effect of our planet which causes a grave environmental concern. This was first indicated in a paper of Arrhenius in 1898. The warming trend of our earth can be evaluated only over longer time periods, but there is a relationship between increasing CO_2 content of the atmosphere and the temperature. (Figures 3 and 4)

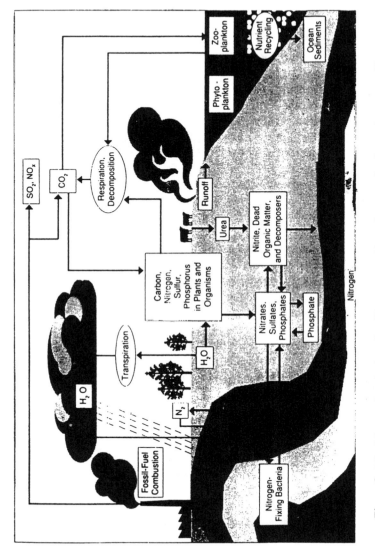

Figure 2. **The biogeochemical cycles. Movement of key elements (carbon, nitrogen, sulfur, phosphorus, and others) through the earth system is critical to the maintenance of life. (from** *Earth System Science: A Closer View,* **report of the Earth System Sciences Committee/NASA Advisory Council, 1988.)**

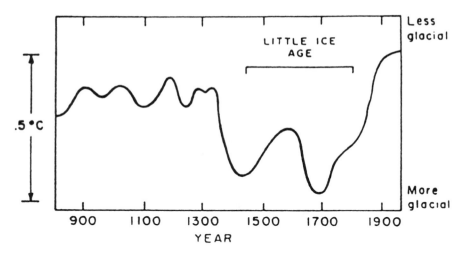

Figure 3. Estimate of the changes in temperature in Europe over the past 1,000 years.

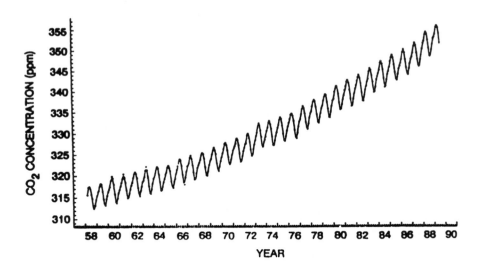

Figure 4. Concentration of atmospheric carbon dioxide – 1958 to 1989 at Mauna Loa Observatory, Hawaii. The dots indicate monthly average concentration.

Recycling carbon dioxide into useful fuels thus would not only help alleviate the question of our diminishing fuel resources, but would at the same time help mitigate the global warming problem.

Fuel Cells. A highly efficient way to produce electricity directly from fuels is effected in fuel cells via their catalytic chemical oxidation.

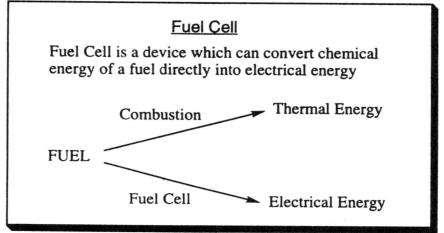

Whereas the principal was known for a long time, large scale practical use is still under development. In its usual application hydrogen and oxygen gases are burned in an electrochemical device producing water and electricity. (Fig. 5)

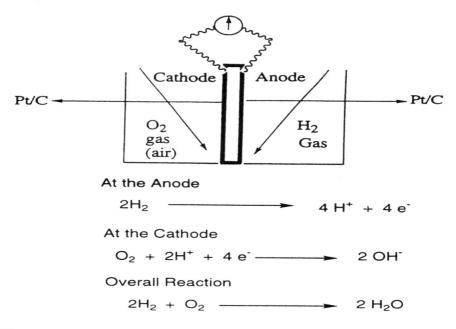

Figure 5. Hydrogen-oxygen fuel cell.

The process is clean, but handling hydrogen and oxygen gas is not only technically difficult but also dangerous. The use of fuel cells, however, is gaining application in static installations or in specific cases, such as space vehicles. Hydrogen gas can also be produced from hydrocarbon sources using reformers which convert them to a mixture of hydrogen and carbon monoxide, which are then separated. Hydrogen burning of fuel cells are, however, limited in their applicability. In contrast, a new approach uses directly liquid fuels, such as methyl alcohol (or its derivatives). Such a direct oxidation liquid feed methyl alcohol based fuel cell was developed in our cooperative effort with Caltech-Jet Propulsion Laboratory, which built all the fuel cells for the US space program. It reacts methyl alcohol with oxygen or air over a suitable metal catalyst producing electricity while forming CO_2 and H_2O. (2) (Fig. 6)

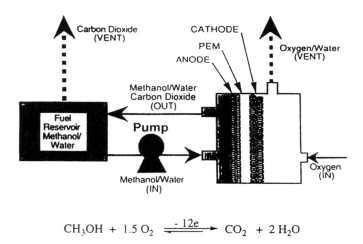

$$CH_3OH + 1.5\,O_2 \underset{\longrightarrow}{\overset{-12e}{\rightleftharpoons}} CO_2 + 2\,H_2O$$

Figure 6. Diagram of liquid feed direct methanol fuel cell.

Recently it was also found that the process can be reversed. Methyl alcohol or related oxygenates can be made from carbon dioxide via aqueous electrocatalytic reduction without prior electrolysis of water to produce hydrogen in what is termed a "reversed fuel cell". This process can convert CO_2 and H_2O electrocatalytically into oxygenated fuels, i.e. formic acid or its derivatives, even methyl alcohol depending on the cell potential used in the fuel cell in its reversed operation. (*3*) (Fig. 7)

The reversed fuel cell accomplishes the electrocatalytic reduction of CO_2 outside the potential range of the conventional electrolysis of water. In its reversed mode the fuel cell is charged with electricity and produces oxygenated methane derivatives such as methyl alcohol, dimethyl ether, dimethoxymethane, trimethoxymethane, trioxymethylene, dimethyl carbonate, methyl formate and the like from carbon dioxide in aqueous solution. The fuel cell thus acts as a reversible storage device for electric power, much more effectively than any known battery. Further recycling of carbon dioxide provides not only the regeneration of fuels but at the same time helps to diminish the atmospheric build-up of carbon dioxide, the most harmful green house gas.

Recycling of CO_2 into CH_3OH or dimethyl ether can subsequently also be used to make, via catalytic conversion, ethylene as well as propylene. These allow ready preparation of gasoline range or aromatic hydrocarbons, as well as the whole wide variety of other hydrocarbons and their derivatives on which we rely in our everyday life. (*4*)

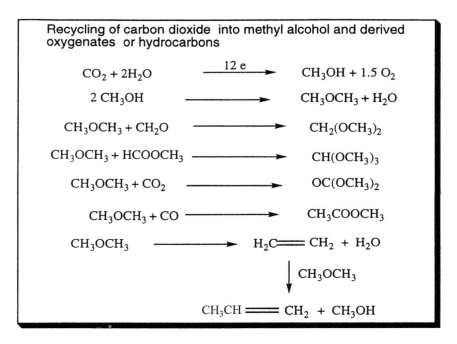

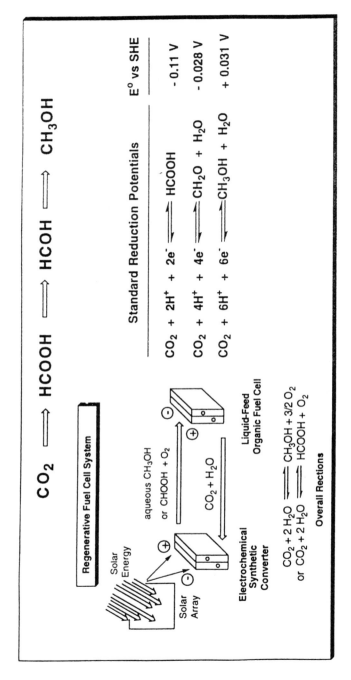

Figure 7.

Decreasing Atmospheric CO_2 Levels: The Need for New Technology

The discussed approach describes a new way to produce hydrocarbons with simultaneously mitigating the environmental carbon dioxide build-up. Green house warming of our planet by excessive burning of fossil fuels producing carbon dioxide is considered such a serious problem that in December 1997 the Kyoto Conference will enact an Agreement establishing and dividing carbon rights. Each country will have an agreed upon quota of its combined fossil fuel allotment which it will be able to burn. Excessive use will be forbidden, necessitating either expensive ways to remove carbon dioxide or using alternative energy sources or atomic energy. Nations will also be able to trade their carbon quotas. Poorer nations thus could sell their allotment to richer industrialized countries. However, we are all citizens of the same planet earth. Whoever burns fossil fuels will put carbon dioxide into the atmosphere which is common to all of us. This agreement, although it may have noble goals and some beneficial effects, can not be considered *per se* as a solution and will not necessarily significantly decrease atmospheric CO_2 levels. Recycling carbon dioxide emissions, in contrast, while producing new hydrocarbon fuels can remove excess harmful carbon dioxide from the atmosphere. I believe there is great promise in this approach. Clearly much must be done to move ahead. The basic science, however, which is the foundation for new technology, is well on its way to be developed.

In summary,

1. The World's population in two centuries, since the industrial revolution, increased nearly seven fold reaching 6 billion. In a few decades it will reach 10 billion. This puts great pressure on our limited resources and the global environment.

2. For human life, besides food, shelter, etc., energy is essential. Our non-renewable fossil fuel (oil, gas, coal) reserves will be seriously depleted during the 21st Century. Besides alternative energy sources, atomic energy *albeit* made safer and solving the radioactive waste disposal problem will inevitably be used on a massive scale.

3. Continued burning of fossil fuels composed of hydrocarbons generates excess carbon dioxide, which contributes to global warming (green house effect). At the same time, with diminishing resources, the need to produce synthetic hydrocarbons as fuels and raw materials for man-made products will necessitate the use of a renewable carbon source. Nature recycles carbon dioxide via photosynthesis in plant life. Man will be able to supplement nature and recycle carbon dioxide into hydrocarbons and their derivatives using water as the source for needed hydrogen.

4. The use of efficient direct oxidation fuel cells to convert liquid fuels (such as methyl alcohol and its derivatives) into electrical energy offers new vistas. At the same time, these devices can also be used for regenerative energy storage by recycling carbon dioxide into fuels while mitigating the global warming effect of carbon dioxide.

Literature Cited

1. Olah, G. A., Molnar, A. *Hydrocarbon Chemistry*, Wiley-Interscience Publishers, New York: 1995; and references therein.
2. Olah, G. A. et al. US Patent 5,559,638, 1997
3. Olah, G. A., Prakash, G. K. S. Provisional US Patent application, May 7, 1997.
4. Olah, G. A. *Acc. Chem. Res.* **1987**, *20*, pp 441 and references therein.

Knowledge-Driven Research:
Benefits to Global Health and Medicine

5

Chemistry of the 20th Century and Beyond: The Structure–Function Relationship

William N. Lipscomb, Jr.

Department of Chemistry and Chemical Biology
Harvard University
12 Oxford Street, Cambridge, MA 02138

In 20th century chemistry a dominant theme has been the relationship between molecular structures and their function. As methods for elucidating structures developed, the structural basis of function became an essential part of ever more complex systems, including complexes relating to living systems. This development is exemplified by the fundamental research of Linus Pauling, followed by a few examples from my basic research. Brief references to other pure research follows. Disturbing trends are seen in the balance between research and applied research in universities, industry and governmental laboratories. Finally, the role of the real seeds of new science (curiosity and idea driven research) is examined, with reference to trends toward short term results and benefits, as now required by recent congressional legislation.

The Structure-Function Relationship

Chemistry includes the study of molecules, their interactions and their transformations. More than any other science, chemistry creates its own new materials, most of which are synthesized using a logic that transcends the relationship of chemistry to other sciences (R. B. Woodward, E. J. Corey). The power of these methods, reinforced by structural studies, has moved chemistry into new areas: mineralogy, physics, materials science and especially biology and medicine. Indeed, the molecular basis of living systems presents a major challenge for the 21st century.

The chemist now thinks of molecules and how they are transformed in terms of the three dimensional structure, and how structure influences changes which occur in nuclear positions and electron distribution as reactions proceed. In this century physicists developed such methods as X-ray diffraction (M. von Laue, W. L. Bragg, W. H. Bragg) and nuclear magnetic resonance (I. Rabi, F. Bloch, E. M. Purcell) for establishing crystal and molecular structures. At first comparatively simple structures were solved, but further developments led gradually to the most complex structures. For examples, L. Pauling formulated rules for deducing structures of minerals, J. M. Bijvoet used multiple heavy atoms in a molecule to solve the structure of strychnine, M. F. Perutz developed this heavy atom method for protein structures, and J. C. Kendrew obtained in 1959 the first structure of a globular protein (myoglobin). Shortly thereafter, Perutz provided the structural basis for control of function (oxygen affinity and regulation) in the hemoglobin molecule.

Pauling went on to many other areas: he, R. B. Corey and H. R. Branson described the helical and sheet structures of proteins in 1951. Pauling dominated in structural and chemical bonding studies, and intermetallic materials. He also contributed to the expansion of chemistry into the then new field of molecular biology. Here, his contributions included the identification of sickle cell anemia as the first example of a molecular disease, complementary structures in transition states of enzyme reactions, and the evolutionary clock at the molecular level. Although his proposal for the structure of DNA was incorrect, his principles of structure were to influence J. D. Watson and F. H. C. Crick in their discovery of the correct structure of DNA in 1953.

When I was at Caltech in the Physics Department (1941), Pauling inspired me to return to the field of chemistry, and became my research advisor for Ph. D. studies. Although most of my efforts of the next five years were on war projects, Pauling remained an inspirational leader both in science and in the war effort. I did not believe all of his ideas, especially on chemical bonding in compounds which have fewer electrons than is required for conventional single, double and triple bonds. In 1946 I began a program (which Pauling thought was not very interesting) at the University of Minnesota in studies of structure, bonding and chemical behavior in the then mysterious boron hydrides, an area which grew into a substantial field of chemistry. The first effort was to establish the three dimensional structures of these compounds by X-ray scattering from single crystals, and then to relate these structures to the chemical behavior of these compounds. All of the boron hydride structures which were then available from electron scattering of gaseous molecules were wrong, and hence Paulings' explanations of molecular properties were also wrong. The unambiguous structures from our research, the description of new types of chemical bonding by us (and by H. C. Longuet-Higgins), and the subsequent severe tests by us of this bonding led to a deep understanding of the chemistry, and to my 1976 Nobel Prize in Chemistry.

Examples: Pure Research to Benefits

Boron Compounds. Some of these boron compounds have been employed to treat brain tumors by neutron activation of radioactivity in boron nuclei. This is my first example of beneficial results arising from pure research which did not have a practical objective as a goal. The orientation of research supported by Federal Funds toward practical goals, initiated by President Johnson, led to the redirection by Oren Williams of the Inorganic Division of the National Science Foundation toward catalysis by transition metals. Support for most of inorganic chemistry of the main group elements was dropped. In particular, my NSF grant was terminated in 1964. Only in recent years has support of most of the main group elements been revived by NSF. Meanwhile, the inorganic chemists of other countries, especially Germany, clearly demonstrated the importance of main group chemistry. Of course, I had to move on to the structure-function relationship in other areas: large organic molecules (e. g. vincristine) and enzymes such as CPA (carboxypeptidase A) and FBPase (fructose-1,6-bisphosphatase), supported by the National Institutes of Health (NIH).

Cancer Chemotherapy. For vincristine, which has been used as one of several pharmaceuticals for successful treatment of childhood leukemia, our structural study established the three-dimensional atomic arrangement which was previously incorrect. Synthesis of this compound (by G. Büchi) then became possible.

Hypertension. For CPA, our interest was and still is how this digestive enzyme cleaves peptide bonds in proteins. The three-dimensional structure severely limits the choices and sequence of the chemical steps, and shows how the zinc ion at the active site aids catalysis of peptide bond cleavage. Studies of enzyme mechanism (the parts of the enzyme directly involved in catalysis, and the sequence of chemical steps) are greatly facilitated by the design and synthesis of small molecules that bind to the active site and block the activity (*1*).

Another enzyme which has a zinc ion bound to the active site is angiotension-converting enzyme (ACE), which cleaves a dipeptide from the decapeptide angiotension I to yield the octapeptide angiotension II. Now angiotension II raises blood pressure, whereas angiotension I does not. Hence, an inhibitor of ACE would be a pharmaceutical which would control hypertension. Such an inhibitor was designed by Ondetti, Rubin and Cushman of Squibb in 1977 (*2*). However, the three-dimensional structure of ACE is not known, and hence our enzyme structure for CPA was used as a surrogate. This was the first example of the use of an enzyme structure for design of a pharmaceutical, a method which is now in wide use. It is called "rational drug design", a phrase which does not necessarily reflect on previously used methods!

The point is that our interest in how an enzyme-bound zinc ion promotes catalysis led to the design of an important pharmaceutical. This result could not have been foreseen, nor would any pharmaceutical company have supported our initial structural studies at that early stage of interest in fundamental science.

Type II Diabetes. FBPase is a critical regulated enzyme in the biosynthetic pathway that makes glucose, our interest was and still is how the enzyme is regulated. AMP (adenosine monophosphate) binds to a regulatory site some distance from the active site, and induces a conformational change that greatly reduces the activity of the enzyme. It is known that FBPase is present in elevated amounts in Type II diabetics. Hence the design of a molecule similar to AMP, but modified to be highly specific and very strongly bound to FBPase, might be useful in treatment of Type II diabetes. Our research group has been working with Gensia Pharmaceuticals for several years to develop such a compound. When we began the study of the structural basis for regulation of FBPase, we knew that this enzyme is in the pathway to make glucose, although we did not then know that the concentration of this enzyme is higher in Type II diabetics than in healthy subjects. Although part of a MERIT AWARD, these studies of FBPase were subsequently refused support three consecutive times plus one appeal, before renewal by the NIH Council.

Fundamental Research: a Few Other Areas. The formation and decomposition of ozone, and the role of human influence was the product of research in universities in the U.S.A. (M. J. Molina, F. S. Rowland, H. S. Johnston, J. S. Anderson) and the Max Planck Institute in Germany (P. J. Crutzen).

Nanotechnology based on the fullerenes arose from fundamental research at universities (R. F. Curl, Jr., H. W. Kroto and R. E. Smalley).

The prostaglandins, and later the related leucotrienes, were synthesized by E. J. Corey, the first completed in 1968. These compounds function in the regulation of blood pressure, blood coagulation, the immune response, childbirth contractions, and other critical processes. In 1969 synthetic routes became available for known prostaglandins, and are used in production by industry and in medical research.

The research which led to NGF (nerve growth factor) was begun in 1947 by Rita Levi-Montalcini. Stanley Cohen joined the project in 1953, and together they shared the Nobel Prize in Medicine in 1986. The amino acid sequence was established in 1971 by Ruth Angeletti and Ralph Bradshaw. The function is to allow nerve cells in culture to behave like sympathetic nerve cells; elimination of NGF causes reversion to the glandular strain (P. Calissano, and also L. Green and A. Tishler). The facinating story as told by Levi-Montalcini is strongly recommended (*3*). Aspects of the detailed processes of function, and especially differentiation, of nerve cells have become clear as a result of this research, and pathological conditions associated with deficiency of NGF or need for extra NGF may now be studied.

The 1945 Nobel Prize in Medicine to A. Fleming, H. W. Florey and E. B. Chain for the discovery of penicillin by Fleming in 1928 and its curative effects in various infective diseases recognized their seminal beginning of antibiotics in chemotherapy (*4*). Florey and Chain were joined by Robert Robinson, one of the few greatest synthetic organic chemists of all time, in a collaboration on the synthesis of penicillin. However, the correct structure, on which the synthesis had to be based, was actually established in an X-ray diffraction study of crystals of penicillin chloride in 1945 by D. C. Hodgkin, who with her research group subsequently established structures for cephalosporin C, vitamin B_{12} and insulin. The function of penicillin is to block a critical step in the synthesis of the cell wall in bacteria, by binding effectively irreversibly to an enzyme in this synthesis (*5, 6*). Not until the very late 1930's was the interest of the pharmaceutical industry awakened, and even then as part of efforts relating to World War II (*7*).

Most of the origins of the biotechnology industries are from pure research in universities, medical schools and research institutes, including the discovery and applications of restriction enzymes that allow precise cleavage of DNA (W. Arber, D. Nathans and H. O. Smith); for preparation of recombinant DNA (P. Berg); for determination of base sequences in DNA and RNA (W. Gilbert and F. Sanger); and for the synthesis of gene fragments (H. G. Khorana).

Five years after C. J. Pedersen joined the DuPont Company in 1957 as a research chemist, he isolated a 0.4% yield of crystals of a ether-type substance that bound sodium or potassium ions so that the resulting complex was soluble in organic solvents. This remarkable substance was the first crown ether (dibenzo-18-crown-6). By 1957 Pedersen described some 50 different crown ethers having hole sizes that would bind various metal ions (8). The detailed structures of some of these molecules were obtained by Mary Truter who confirmed the proposed structures (9). Pederson's discoveries opened up new directions, including J.-M. Lehn's cryptates, D. J. Cram's host-guest chemistry, and C. J. Liotta's phase-transfer catalysis. In 1987, Cram, Lehn and Pederson shared the Nobel Prize in Chemistry, with special reference to structure-specific interactions of high selectivity.

D. R. Hershbach, in a Symposium (10) on the *Flight from Science and Reason* traces a chain of fundamental discoveries from atomic beams (O. Stern, 1924), to nuclear magnetic resonance (NMR or MR) in atomic beams (I. Rabi, late 1930's) to NMR of nuclei in solutions of molecules (E. Purcell, F. Bloch, 1945) to MR for imaging in medicine (The word nuclear is unpopular with patients). The interaction with light in these experiments gave rise to the LASER (C. Townes and N. Bloembergen). Also chemistry at its most elementary steps was divulged in crossed beams; and the fullerenes (C_{60} and beyond) were discovered and being made into very unusual "nanomaterials" and even being tested after chemical modification as potential inhibitors of the AIDS virus. Perhaps the most extraordinary influence of Stern's original experiment with beams of silver atoms is that the result after passage of a beam through an inhomogeneous magnetic field only two beams were formed, not the continuum of beams expected by all physicists of that early time. Clearly a new comprehensive quantum mechanics was made necessary, social constructionists notwithstanding. The question for future discoveries of this magnitude, is, "Will we maintain a structure of research support which will foster fundamental discoveries that, *in an initially unforeseen way*, give rise to later stages of discoveries comparable with these discoveries of lasers or magnetic resonance?"

The Importance of Basic Research

Basic Research Versus Accountable Research.   So far, I have tried to describe basic (pure, fundamental) research, rather than define it. For examples, the scientist may perceive how to resolve an anomaly, an unsolved problem, or a conflict among results from other scientists. A favorite area is to explore a new interface between existing fields, especially as opened up by a new idea or a new experimental method. There are no rules (7, 11). Basic research is often driven by the interest and curiosity of an investigator. The initial proposal or idea may seem difficult or improbable at the outset, and the initial stages can be messy or seem disorganized. Indeed, a new line of research that changes science (one of the criteria sought by Nobel Prize Committees) is often greeted by colleagues or referees as a (a) wrong, (b) not worth doing, (c) has already been done, or (d) impossible to do. Moreover, a new basic research area may take many years to achieve an identity. Frequently the result is failure. The counterforces against basic research are many, frequently including rejection of support by granting agencies.

Accountable (applied) research ordinarily has defined goals, and usually has constraints that require results in a short time.

Universities. Even in universities there are pressures for short term results arising from the agencies that support research (discussed below). Such research is usually carried out in groups of graduate students and postdoctoral fellows. The pressures also arise in part because of the limited time of a few years that these individual coworkers stay. Long term projects must therefore be divisible into units suitable for a Ph. D. thesis or a postdoctoral stay of a few years. Moreover, there are pressures from both coworkers and granting agencies for publications.

It could be argued that basic research should be done in research institutes, such as the Max Planck Institutes in Germany. For example, H. Michel's long term research on crystallization of membrane proteins, done in an MPI, probably could not have been carried out in the USA. On the other hand, research institutes (unless they are closely allied with a university) do not produce new generations of scientists, and they may draw the best teacher-research talent away from universities.

Industries. Support for basic research in industry has been cut sharply in the 1990's as a result of competition, downsizing, budget balancing, shorter horizons (3 years or less) increased emphasis on process development and less governmental support. Deregulation has also played a role, as illustrated by the Bell Telephone dismemberment. In chemistry, the pharmaceutical companies are hiring, and some basic research is carried out through agreements between companies, particularly as large companies enter partnerships with smaller companies on specific developments. It remains to be shown that innovation in small companies compensates for the decline of basic research (as opposed to development) in the established major companies. Nor is it clear that governmental support of basic research in universities and research institutes will be sufficient to maintain the leadership of the USA, in that innovative research which leads to later development.

Congress. In the government, the drive for accountability, started in earnest in the early 1960's, has further tipped the balance away from long-term research to shorter terms. The "competitive crises" of the early 1980's resulted in the ORTAs (Offices of Research and Technology Applications) and the CRADAs (Cooperative Research and Development Agreements) which provided support for partnerships between research laboratories and industry. These laboratories, particularly those of the Government R+D, were required to show that their research had a favorable impact on the economy. The drive toward balancing the Federal Budget led Congress in the early 1990's to require that laboratory budgets (especially NASA and the Department of Energy) show in dollar amounts and job creation the influence on the economy (*12*). The GPRA act   (Government Performance and Results Act of 1993) requires documentation of performance as measured by effective results. The objectives of this act are "to improve the efficiency and effectiveness of federal programs by establishing a system to set goals for program performance and to measure results". This act of Congress (*13*) is incompatible with basic research, and may well cost the United States its leadership in science,   not immediately, but progressively and noticeably within a decade or so.

The Granting Agencies. Although GPRA contains a provision to allow an agency to request funding for proposals for which it is "infeasible or impractical to express a performance goal in any form for the program activity", the permission of OMB is required. For reasons which may involve OMB officials and congressional staff, the NIH and NSF have not explored use of this provision. Indeed, basic research fits this provision very well.

Although short term immediately useful and practical results may not be the stated agenda of NIH and NSF, there is a recent development of such policies in the comments of referees to grant proposals. Moreover, the selection of referees needs a quality upgrade, and particularly enough care so that their comments address the science (not just the techniques), and do not contain outright lies. In simple terms, the greatest emphasis for choice of referees should be on originality and competence. Also, reorganization of the granting divisions based on the scientific goals rather than on techniques (*14*) would be a much needed refocus, and would, for example, restore the appropriate balance toward structure and function, not just structure.

Beyond 2000. The critical parts of this paper do not reflect pessimism of this author about the future. Rather, these comments are meant to be helpful in catalyzing changes. I am a firm believer in the indispensable value of basic research both as a human endeavor and as a source of presently unknown science and discovery for the benefit of mankind. No mission oriented research can be a substitute, although it too is valuable. The past is our best guide: in the medical sciences the major advances have been based on investigator-driven curiosity, usually by chemists, biologists or physicists. Their discoveries, for examples, of vaccines, antibiotics, genetic engineering and X-rays, were initially unrelated to the drugs and instruments which followed. These initial long term research programs, unsupportable by industry, yielded extraordinary benefits (*15*).

Finally, I certainly hope that my essay does not discourage younger people from entering science as a profession, and I remain certain that a very large number of as yet unknown benefits will arise in unforeseen ways from basic research, and in foreseen ways from applied research.

Acknowledgement: The experimental studies in proteins reported here from my research were supported by NIH grant GM06920.

Appendix: Chemistry of the 20th Century. In the discussion following the presentations, a question was asked by someone in the audience about an introductory chemistry course different from the standard first year university course.

In the period 1962-1968 several members of our Department felt that the then standard undergraduate courses for the first three years were not sufficiently challenging for our more advanced undergraduates, and were in need of updated changes. A four semester sequence was founded, taught in order by W. N. Lipscomb, E. J. Corey, F. H. Westheimer and E. B. Wilson, Jr. High school chemistry and physics was a prerequisite. Starting with the goal of understanding the periodic table, valences and chemical bonding, I began with semiquantitative emphasis on the ideas of the wave properties of matter, and proceeded to transition metal chemistry and an introduction to phenomenological thermodynamics. The modernization of organic synthesis by Corey, the enlargement toward biochemistry by Westheimer, and the thermodynamic-statistical mechanical approach to chemistry by Wilson then brought this group of students in two years to the level that they were able to take graduate courses and do undergraduate research. The laboratory associated with the course was planned to carry through synthesis, analysis (plus spectroscopic and NMR studies), and reactivity-kinetics studies on the same material, when feasible. The very first laboratory experiment addressed the eternal questions of beginning students about three-dimensional standing waves (orbitals). As an illustrative model the students were given a signal generator of variable frequency, a loudspeaker, a stethoscope, and an empty box approximately 16"X20"X24". The generator was placed at the center of the box, and nodes were found at the various resonance frequencies. It worked: no more questions about 3D standing waves!

The second introductory course, called Chemistry of the 20th Century, was available only for non-science majors (1980-1990). I decided to create this new kind of introductory chemistry partly because there was a special need for a course for non-scientists, and because chemistry is closely related to areas such as the individual's well-being, the environment, resources, and energy for the long term future; in addition to these areas chemistry is rapidly expanding into materials science and molecular biology.

In this new course there was no laboratory although some demonstration experiments were done. Each student was required to submit a paper of about 10 pages on a suitable topic not covered in the lectures. There was a problem set each week, and a midterm and a final written examination. Although there was no text, extensive material was distributed based on articles from sources such as Scientific American, Science, Nature, Chemical and Engineering News, and my own notes, including prints of the transparencies used in the lectures. The topics are listed below, for the individual lectures which included, in part, the beginnings of chemistry's movement into molecular biology. Also, there was emphasis on environmental, energy and health issues of interest to most students. Within this framework it was possible to include many of the ideas of chemistry, several unresolved issues, and an appreciation of the beauty of some of the counterintuitive results of the properties of matter on the atomic and molecular scale.

1. Atoms and the periodic table
2. Matter waves. The H atom (uncertainty principle)
3. The periodic table. Ionic valences. Transition metals. Rare earths
4. The covalent bond in H_2. Electronegativities
5. pH. Acid rain. SO_2, NO_x
6. Reactions in the troposphere. Organic nomenclature
7. Weak acids, buffered solutions
8. Review buffered solutions
9. Ozone layer I. The three catalytic cycles of the O_3 layer
10. Ozone layer II. The anarctic hole. The gas laws
11. Nobel prizes in science for the year
12. The greenhouse effect
13. Introduce organic chemistry. D and L amino acids
14. Organic synthesis of amino acids. The 21 coded amino acids
15. Primary protein structure, plus α helix, sheets. Hemoglobin secondary, tertiary, and quaternary protein structures
16. Carboxypeptidase A mechanism
17. DNA→RNA→protein
18. DNA, RNA, protein biosynthesis, ribosome
19. The genetic code. Protein biosynthesis
20. Base sequencing
21. DNA replication, restriction enzymes. recombinant DNA

22. Steps in recombinant DNA, containment levels. Antibody resistant plasmids
23. Moveable pieces of DNA, transposons
24. Oncogenes. antioncogenes
25. AIDS. Molecular targets for inhibition
26. Cancer epidemiology
27. Cancer survival statistics. Mutations. Benzopyrene
28. Bhopal. Dioxin. Bendectin
29. Energy sources, health effects, oil and gas
30. Atomic energy. Packing fraction, OKLO reactor. Storage problems
31. Safe reactors
32. Other energy sources

Literature Cited

1. Lipscomb, W. N.; Sträter, N. *Chem. Rev.* **1997**, *96*, pp 2375-2433.
2. Ondetti, M. A.; Rubin, B.; Cushman, D. W. *Science* **1977**, *196*, pp 441-444.
3. Levi-Montalcini *In Praise of Imperfection,* Basic Books, New York, NY 1988, pp 123-168.
4. Fleming, A. *Brit. J. Exp. Pathol* **1929**, *10* pp 226-236.
5. Tipper, D. J., and Strominger, J. L. **1965**, *Proc. Natl. Acad. Sci. USA 54,* pp 1133-1141.
6. Boyd, D. B. *Proc. Natl. Acad. Sci. USA* **1977**, *74* pp 5239-5243.
7. Medawar, P. B. *The Limits of Science,* Harper and Row, New York, NY 1984, pp 45-54.
8. Pedersen, C. *J. Am. Chem. Soc.***1967**, *89* pp 2495 and 7017.
9. Truter, M. R. and Pederson, C. *Endeavour* **1971**, *30,* p 4278.
10. Herschbach, D. R. *Ann. New York Acad. Sci* **1996**, *775*, pp 11-30.
11. Lipscomb, W. N. *Aesthetic Aspects of Science in the Aesthetic Dimension of Science,* D. W. Curtin, ed. *The Philosophical Library*, New York, NY 1982, pp 1-124.
12. Papadakis, M. *The Scientist* October 27, 1997, p 8.
13. Barnes, D. M. *J. NIH Research* **1995**, *7*, p 10.
14. Staddon, J. E. R. *J. NIH Research* **1997**, *9*, p 13.
15. Kornberg, A. *Science* **1996**, *273* August 16 pp 855-860.

6

Organic Synthesis: From Art and Science to Enabling Technology for Biology and Medicine

K. C. Nicolaou and Janet L. Gunzer

Department of Chemistry and The Skaggs Institute for Chemical Biology
The Scripps Research Institute
10550 North Torrey Pines Road, La Jolla, CA 92037

Department of Chemistry and Biochemistry
University of California, San Diego
9500 Gilman Drive, La Jolla, CA 92093

Since ancient times, humankind recognized particular changes in matter under certain conditions and was quick to exploit them. Cooking, wine-making, and alchemy are examples of processes in which matter changes from one form to another. These transformations of matter were practiced for ages without an understanding of the science involved. It was only in the nineteenth century that we began to comprehend such changes on the molecular level, and chemists soon realized that they could create new substances by rational scientific methods. Today, this awesome power, known as synthetic chemistry, promises to bring even more drastic changes to our lives than the ones we already experience in our everyday existence: foods, medicines, shelter, cosmetics and high tech materials for aircraft, automobiles, computers, and other electronic devices are some of the ways chemistry improves human well-being and drives progress. Organic synthesis, in particular, currently serves as the enabling technology for biology and medicine. It is due to the basic advances in this field that we are able to understand a great deal about life and to rationally design and develop new medicines for people. In this article, we will examine the current status of organic synthesis as exemplified by projects from our laboratory and attempt to look ahead to the forces that will drive it to even higher levels of sophistication and performance as a tool for biology and medicine.

Introduction

Life on Earth is a consequence of molecules, for the most part organic, interacting with each other in a harmonious way. The dynamic nature of living systems requires continuous change in the constitution of their components, which are brought about in a most admirable manner at ambient temperatures, aqueous media and with very little waste. Even though humankind observed and practiced such matter-changing processes for centuries, it was only recently that the art and science of organic synthesis — the construction of carbon compounds from other readily available substances — was placed on a rational and systematic footing. The year 1828 marks the beginning of organic synthesis with Wöhler's transformation of ammonium cyanate [NH_4CNO] to urea [$CO(NH_2)_2$]. Since then, impressive strides have been made and the science of organic synthesis emerged as one of the most powerful and significant scientific developments of the 20th century.[1] This truth can be fully appreciated by reflecting on the ways that chemical synthesis has changed and shaped our world[2] by providing us new foods, medicines, shelter, cosmetics, agricultural products, and high tech materials — so many of the things that we see, touch and smell in our everyday lives. The significance and impact of chemical synthesis on our lives is often missed by most lay people due to the lack of knowledge and proper attribution in the media. But the coverage is there, disguised in news about medical breakthroughs (drug discovery and development), space exploration, transportation and communication (high tech materials for spaceships, aircraft, automobiles, computers, electronic devices), fashion (new fabrics, dyes, jewelry, fragrances), nutritional products (foods, vitamins, hormones) and toys, household gadgets and construction materials (polymers, composites, paints). Society has come to rely more and more on synthetic materials as opposed to the naturally derived ones due to our acquired power to create new substances in the laboratory and choose the ones with the right properties for the task. Of the known organic compounds, more than 90% are synthetic. Even though it is true that nature still holds most of its molecular diversity secret from us, it is also certain that we have to our disposal the power and potential to outdo nature in the pursuit of structural diversity by virtue of our ability to rapidly generate, in a combinatorial fashion, astronomical numbers of new compounds. This capacity of organic synthesis is, of course, a blessing because serious challenges still remain. For example, we need cures for cancer, AIDS, Alzheimer's disease, obesity, diabetes, arthritis, stroke and cardiovascular disorders and protection against new and dangerous viruses and bacteria, just to mention a few medical problems; high tech materials for construction of faster, stronger and safer transportation vehicles, computers, buildings and engines; and nutritional and healthy foods and vitamins for good health and prevention of disease.

For its pivotal role within chemistry, biology, medicine, and engineering and high technology, chemical synthesis can be considered as central and highly enabling. Constantly enriching and increasing the capabilities of chemical synthesis is, therefore, of extreme importance and we must be vigilant in searching for new ways and means to sharpen its edge and expand its reach. Research in organic synthesis is currently rigorous and vibrant, a healthy sign of its importance and recognition as a crucial discipline. The endeavors of organic synthesis usually fall within the areas of inventing new synthetic technology or target synthesis. Target synthesis may aim at designed target molecules or naturally occurring substances, whereas invention of new synthetic technology refers to devising means for the more efficient construction of organic molecules. These endeavors are directed toward the development of the science of organic synthesis for its own sake and should be differentiated from the use of organic synthesis as a tool in applied situations where the emphasis is placed on the properties and uses of the target molecule, as for example in biology and medicine. In the section below, we will briefly touch upon ways in which the frontiers of organic synthesis are pushed forward, and towards a magic kingdom where molecules can be made at will and at the push of a button in ways rivalling those used by nature.

New Synthetic Methods

Organic synthesis derives its power from synthetic reactions or methods and from synthetic strategies. Methods allow the synthetic chemist to make or break bonds between atoms, operations needed for the construction of molecules. Despite the abundance of such methods, challenges still remain, particularly in terms of selectivity, efficiency, environmentally sound conditions and the ability to generate molecular complexity. Amongst classic synthetic reactions are many that we take for granted today. They include the aldol reaction, Grignard and related organometallic reactions, Diels-Alder cycloaddition, esterification and the reactions for forming peptide and phosphate bonds and glycosidic linkages, Wittig reaction, hydroboration reaction, epoxidation reaction and the photochemical 2+2 cycloaddition reaction. Newer processes such as the various reactions used in asymmetric synthesis, particularly those which are catalytic in nature, have expanded the scope of organic synthesis to include access to enantiomerically enriched compounds. Palladium-catalyzed reactions resulted in numerous new methods for constructing carbon-carbon bonds and a new generation of cyclization reactions involving lactonization, etherification and olefin metathesis have aided the formation of complex and challenging medium and large rings.

The invention of new synthetic reactions is currently one of the most active areas of research. Particularly intense is the activity in asymmetric synthesis, the use of organometallic and biological catalysts, and the generation and trapping of reactive species such as free radicals. Research in this field should be continued in order to achieve higher selectivities and efficiencies, less wasteful procedures, and attainment of higher molecular complexity. Newer synthetic technologies with higher levels of efficiency will allow the production of desired products more economically and within environmentally acceptable controls. Most importantly, such methods will allow the strategist to design viable and practical routes to even more complex and sensitive target molecules. In the next section, we will discuss the interplay between new synthetic technology, strategy and total synthesis.

Strategy and Total Synthesis

Soon after the advent of organic synthesis, natural products became targets for synthetic chemists. Urea (Wöhler, 1828) was followed by acetic acid (Kolbe, 1845), glucose (Fischer, 1890), camphor (Komppa, 1903; Perkin, 1904), α-terpinol (Perkin, 1904), tropinone (Robinson, 1917), haemin (Fischer, 1929), equilenin (Bachman, 1939), pyridoxine hydrochloride (Folkers, 1939) and quinone (Woodward and

Doering, 1944). But it was after World War II that organic synthesis and total synthesis flourished, reaching unprecedented and impressive heights. R. B. Woodward and his school surpassed previous records of complexity and elegance, introducing a touch of artistry to the discipline. Strategy was the prime concern of the Woodward school, whose works are considered classics and are still admired today. Included in Woodward's total syntheses are strychnine (1954), reserpine (1958), cephalosporin (1965) and vitamin B_{12} (1973, with A. Eschenmoser).

The next quantum leap was already in the works from the 1960's when E. J. Corey put forward his principles of strategic design through retrosynthetic analysis. The following decades witnessed higher levels of sophistication with complex structures from several classes of natural products yielding to total synthesis. In addition to strategy, emphasis was now placed on new synthetic methods as part of the total synthesis program. The conceptual advances made by R. B. Woodward and E. J. Corey were rewarded with Nobel Prizes in 1965 and 1990, respectively.

Total Synthesis and Chemical Biology

A new era in natural products synthesis has emerged in the 1990's with a strong identity, that of adding yet another dimension to the total synthesis program, namely chemical biology. Often today, we view the ideal total synthesis program as one directed towards a target with the following characteristics: (1) novel molecular architecture; (2) important biological activity; and (3) interesting mechanism of action. (However, we must not forget the importance of the criterion of sheer challenge!). Most importantly, the opportunity for discovery and invention is considerably enhanced by the incorporation of new synthetic technology and of chemical biology studies as integral parts of the program. It is this philosophy that rewarded us, in particular, with so many findings in new synthetic technology, strategy and biology in pursuit of new targets such as calicheamicin γ^1_1, rapamycin, Taxol™, zaragozic acid A, brevetoxins A and B, epothilones A and B, eleutherobin and sarcodictyin A. These and other projects will be highlighted below with special emphasis on molecular structure, retrosynthetic analysis and synthetic strategy, new synthetic technology and chemical biology. Many beautiful examples of such endeavors adorned the chemical literature and we apologize to their authors for not being able to include them here due to the obvious limitation of space.

Calicheamicin γ_1^I. Isolated from a species of bacteria in the 1980's, calicheamicin γ_1^I (**1**, Figure 1)[3] represents a growing class of antitumor antibiotics known collectively as enediynes. Its stunning molecular architecture is characterized by a conjugated 10-membered ring embedded within a bicyclic system that also carries a 6-membered ring enone and an allylic trisulfide. The enediyne core is joined to a "colorful" oligosaccharide domain that includes, among other features, an NH-O glycoside bond, a thioester, and an iodine atom, all apparently necessary for its potent DNA-cleaving properties and phenomenal antibiotic and antitumor activity.

Calicheamicin's fascinating mechanism of action (see Figure 2, **2→3→4→5**) involves initial nucleophilic attack on the trisulfide moiety generating an internal sulfur nucleophile which engages the enone double bond, resulting in spontaneous Bergman cycloaromatization. The latter reaction leads to a benzenoid diradical (**4**), a

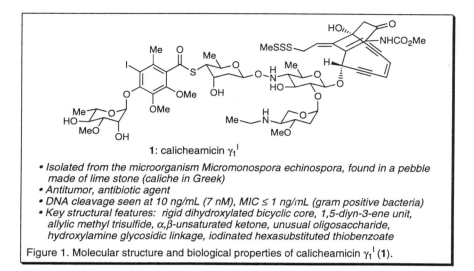

1: calicheamicin γ_1^I

- *Isolated from the microorganism Micromonospora echinospora, found in a pebble made of lime stone (caliche in Greek)*
- *Antitumor, antibiotic agent*
- *DNA cleavage seen at 10 ng/mL (7 nM), MIC ≤ 1 ng/mL (gram positive bacteria)*
- *Key structural features: rigid dihydroxylated bicyclic core, 1,5-diyn-3-ene unit, allylic methyl trisulfide, α,β-unsaturated ketone, unusual oligosaccharide, hydroxylamine glycosidic linkage, iodinated hexasubstituted thiobenzoate*

Figure 1. Molecular structure and biological properties of calicheamicin γ_1^I (**1**).

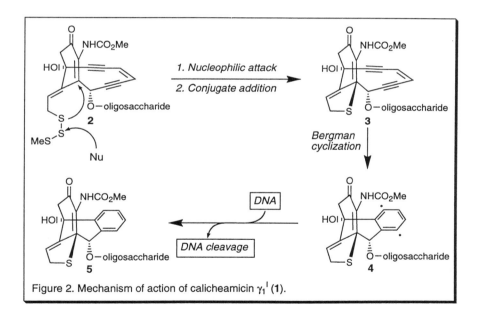

Figure 2. Mechanism of action of calicheamicin γ_1^I (**1**).

reactive species that has the ability to cleave DNA through double strand cuts. The potential of calicheamicin in cancer chemotherapy coupled with its attractive molecular architecture and intriguing mechanism of action offered an irresistible challenge to synthetic chemists and a wonderful opportunity to discover and create new art and science, particularly in chemistry and biology.

From the retrosynthetic point of view, calicheamicin γ_1^I (**1**) renders itself to the strategic bond disconnections indicated in Figure 3. After a campaign of almost five years, calicheamicin γ_1^I (**1**) yielded to total synthesis.[4] The successful strategy involved coupling and elaboration of advanced intermediates **6** and **7** (Figure 3) and features a number of new strategies, including the intramolecular 1,3-dipolar cycloaddition shown in Figure 4 for the construction of the enediyne core. The wealth of new synthetic technology developed during this program can be appreciated in its fullest by reviewing the original publications and review articles describing these accomplishments.[5-7]

The calicheamicin program included a series of chemical biology studies directed at the design, chemical synthesis and chemical and biological investigation of a number of enediynes. Four designed enediynes with interesting reactivity and biological profiles are shown in Figures 5 (**11**: calicheamicin θ_1^I) and 6 (enediynes **12-14**). These studies, which would have not been possible without the enabling power of organic synthesis, enriched our knowledge regarding the fundamental nature of the enediynes, aided in the elucidation of their mechanism of action, and produced a number of potential leads and drug candidates.

Rapamycin.   Rapamycin (**15**, Figure 7) is one of the three most fascinating immunosuppressive agents[8] to come from nature, the other two being FK506 and cyclosporin. Immunosuppressants are the drugs that make organ transplantations possible and allow patients receiving such organs to lead normal lives without complications arising from rejection. Isolated in the early 1970's from soil bacteria,[9] rapamycin joined FK506 in the 1980's as a rival to cyclosporin. Its fascinating story in chemistry, biology and medicine is still evolving. Rapamycin's challenging molecular structure (**15**, Figure 7) coupled with its newly elucidated mechanism of action (Figure 8), fascinated synthetic chemists[10-11] and presented unique opportunities for research combining chemical synthesis and biology.

The strategy for the total synthesis of rapamycin (**15**) in our laboratories was based on the bond disconnections and retrosynthetic analysis shown in Figure 9. The synthetic campaign was accompanied by the development of several new processes and intermediates, the most daring and useful, perhaps being the double Stille coupling based "stitching" cyclization depicted in Figure 10. This reaction produced rapamycin (**15**) from the unprotected divinyliodide **20** in a single step and pointed to a new concept for the construction of macrocycles.

In addition to the development of the new synthetic technology that accompanied this program, an excursion into chemical biology produced RAP-Pa (**21**, Figure 11), a designed analog of rapamycin with only the latter's partial biological action.[12] Lacking rapamycin's FRAP interaction domain, this synthetic compound binds to the FKBP12, but not to FRAP and, therefore, exhibits no immunosuppressive properties. Such synthetic molecules are serving as tools for biology and medicine, underscoring once again the enabling nature of organic synthesis in these disciplines.

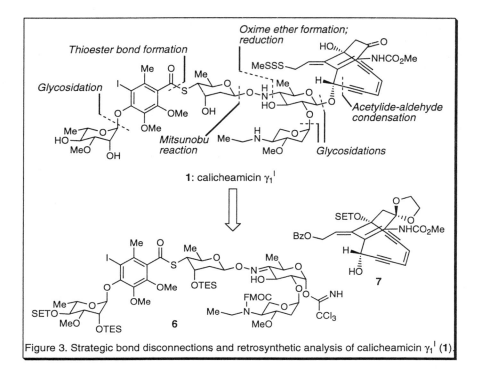

Figure 3. Strategic bond disconnections and retrosynthetic analysis of calicheamicin γ_1^I (1).

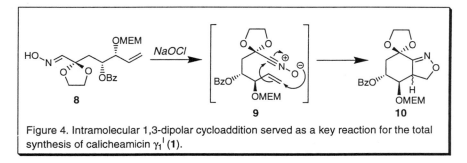

Figure 4. Intramolecular 1,3-dipolar cycloaddition served as a key reaction for the total synthesis of calicheamicin γ_1^I (1).

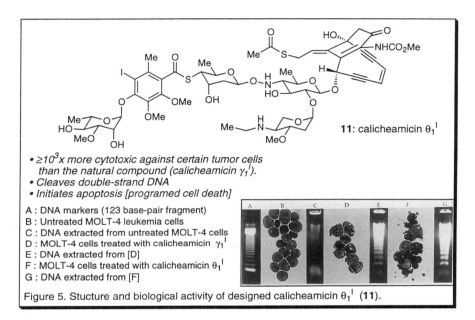

- $\geq 10^3 x$ more cytotoxic against certain tumor cells than the natural compound (calicheamicin γ_1^I).
- Cleaves double-strand DNA
- Initiates apoptosis [programed cell death]

A : DNA markers (123 base-pair fragment)
B : Untreated MOLT-4 leukemia cells
C : DNA extracted from untreated MOLT-4 cells
D : MOLT-4 cells treated with calicheamicin γ_1^I
E : DNA extracted from [D]
F : MOLT-4 cells treated with calicheamicin θ_1^I
G : DNA extracted from [F]

Figure 5. Stucture and biological activity of designed calicheamicin θ_1^I (**11**).

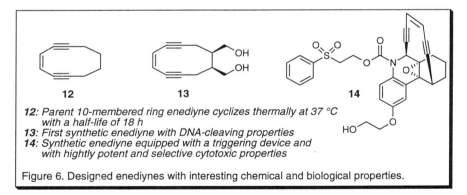

12: Parent 10-membered ring enediyne cyclizes thermally at 37 °C with a half-life of 18 h
13: First synthetic enediyne with DNA-cleaving properties
14: Synthetic enediyne equipped with a triggering device and with hightly potent and selective cytotoxic properties

Figure 6. Designed enediynes with interesting chemical and biological properties.

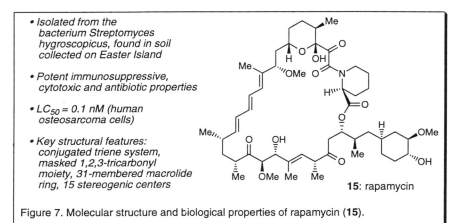

- Isolated from the bacterium *Streptomyces hygroscopicus*, found in soil collected on Easter Island

- Potent immunosuppressive, cytotoxic and antibiotic properties

- $LC_{50} = 0.1$ nM (human osteosarcoma cells)

- Key structural features: conjugated triene system, masked 1,2,3-tricarbonyl moiety, 31-membered macrolide ring, 15 stereogenic centers

15: rapamycin

Figure 7. Molecular structure and biological properties of rapamycin (**15**).

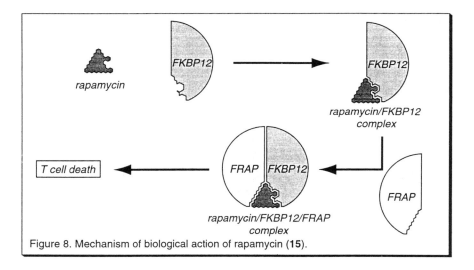

Figure 8. Mechanism of biological action of rapamycin (**15**).

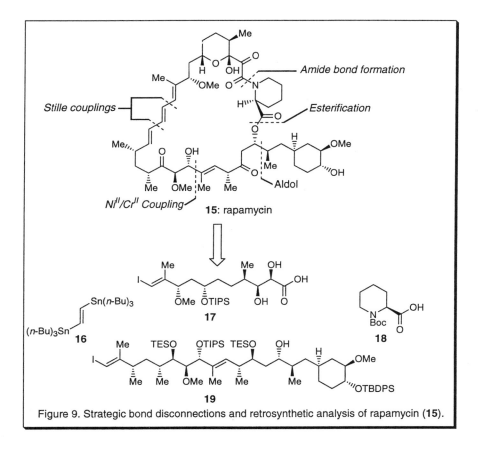

Figure 9. Strategic bond disconnections and retrosynthetic analysis of rapamycin (**15**).

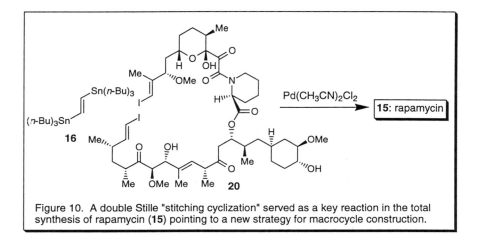

Figure 10. A double Stille "stitching cyclization" served as a key reaction in the total synthesis of rapamycin (15) pointing to a new strategy for macrocycle construction.

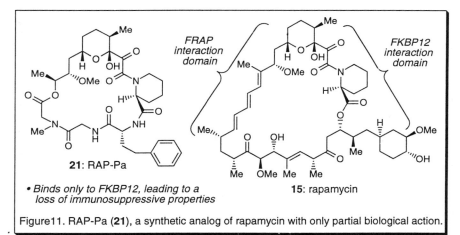

21: RAP-Pa

• *Binds only to FKBP12, leading to a loss of immunosuppressive properties*

15: rapamycin

Figure11. RAP-Pa (21), a synthetic analog of rapamycin with only partial biological action.

Balanol. Isolated from a species of fungi, balanol (**22**, Figure 12) is a potent protein kinase C inhibitor, whose structure is both novel and synthetically challenging. The biological mechanism of action[14] of balanol is thought to involve competitive binding against ATP in the catalytic domain of protein kinases and, therefore, the compound behaves as a modulator of important signal transduction pathways (Figure 13). Inhibition of these pathways may result in beneficial clinical effects against diseases such as cancer, diabetes, inflammation and HIV infection. The perceived high probability of making contributions in chemical synthesis and chemical biology prompted us to initiate a program directed at the total synthesis of balanol.

The general strategy for our total synthesis[15] of balanol (**22**) is shown retrosynthetically in Figure 14, in which key intermediates **23-25** are defined. Of particular interest was the construction of the sterically congested bis(aryl) ketone **23** for which a new method of general applicability was developed (Figure 15, **26→27→28**).

The developed chemistry for the total synthesis of natural balanol (**22**) enabled the chemical synthesis of a series of designed balanols for chemical biology studies.[16] Shown in Figure 16 are two biologically active synthetic balanols (**29** and **30**) with selective inhibitory properties for certain kinases. Such compounds, besides being useful as tools for biology, may have potential in cancer therapy and other disorders.

Zaragozic Acid A (Squalestatin S1). Zaragozic acid (**31**, Figure 17) with its unusual molecular architecture and rich biological action, presented yet another research opportunity for organic synthesis. Isolated from a species of fungi found in soil collected in the Zaragoza province in Spain,[17] this natural product is characterized by an unprecedented molecular structure comprised of a highly oxygenated bicyclic core carrying three carboxylic acid groups and two fatty acid chains. Zaragozic acid's biological properties include cholesterol lowering effects in animals, antifungal activity and farnesyl transferase inhibition. Its cholesterol lowering properties are a consequence of its ability to inhibit squalene synthase (Figure 18), an enzyme involved in the biosynthesis of cholesterol.[18] This observation alone stimulated intense interest in academia and the pharmaceutical industry.

Our successful synthetic strategy towards zaragozic acid A[19] took advantage of the bond disconnections and retrosynthetic analysis depicted in Figure 19. An aesthetically pleasing and highly effective skeletal rearrangement provided a pathway to the core of zaragozic acid A from a readily accessible intermediate as shown in Figure 20 (**35→36→37**). This novel cascade and the rest of the synthetic sequence, which can be found in the original literature,[19] served as the basis for the construction of the designed zaragozic acid analog **38** (Figure 20).

Swinholide A. Isolated[20] from the Red sea sponge Theonella *swinhoei*, swinholide A (**39**, Figure 21) represents an increasing number of macrolide structures with important biological actions. The C2 symmetrical nature of the molecular structure of swinholide A adds to its architectural appeal as a synthetic target. The 44-membered macrocyclic ring together with the four pyran rings and 30 stereogenic centers promised an exciting problem-solving exercise in any campaign for its total synthesis. On the other hand, its intriguing mode of action against fungi and tumor cells by a mechanism involving sequestering of actin dimers and disruption of the actin cytoskeleton, invites chemical biology studies.

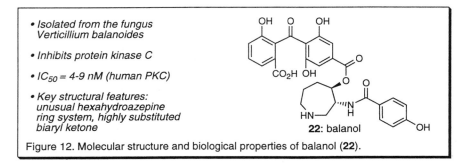

- *Isolated from the fungus Verticillium balanoides*

- *Inhibits protein kinase C*

- IC_{50} = 4-9 nM (human PKC)

- *Key structural features: unusual hexahydroazepine ring system, highly substituted biaryl ketone*

22: balanol

Figure 12. Molecular structure and biological properties of balanol (**22**).

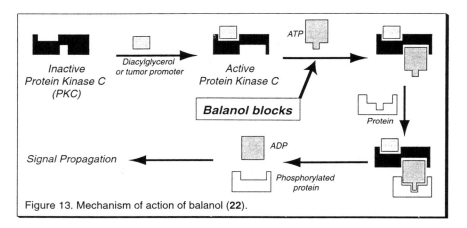

Inactive Protein Kinase C (PKC)

Diacylglycerol or tumor promoter

Active Protein Kinase C

ATP

Balanol blocks

Protein

Signal Propagation

ADP

Phosphorylated protein

Figure 13. Mechanism of action of balanol (**22**).

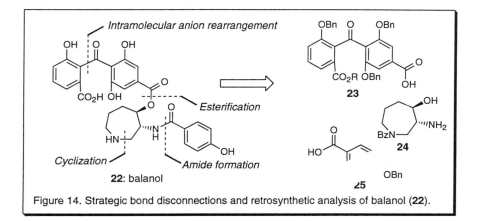

Figure 14. Strategic bond disconnections and retrosynthetic analysis of balanol (**22**).

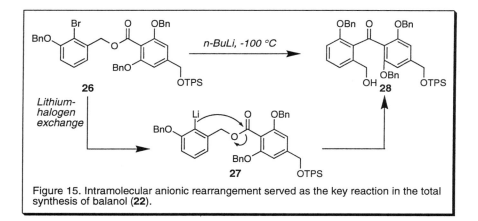

Figure 15. Intramolecular anionic rearrangement served as the key reaction in the total synthesis of balanol (**22**).

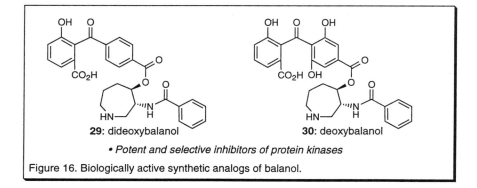

29: dideoxybalanol   **30**: deoxybalanol

• *Potent and selective inhibitors of protein kinases*

Figure 16. Biologically active synthetic analogs of balanol.

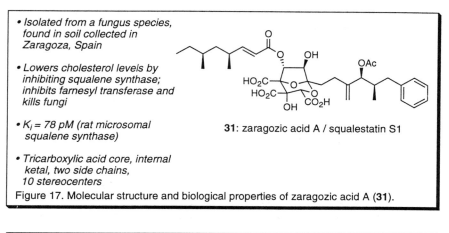

- *Isolated from a fungus species, found in soil collected in Zaragoza, Spain*

- *Lowers cholesterol levels by inhibiting squalene synthase; inhibits farnesyl transferase and kills fungi*

- $K_i = 78\ pM$ *(rat microsomal squalene synthase)*

- *Tricarboxylic acid core, internal ketal, two side chains, 10 stereocenters*

31: zaragozic acid A / squalestatin S1

Figure 17. Molecular structure and biological properties of zaragozic acid A (**31**).

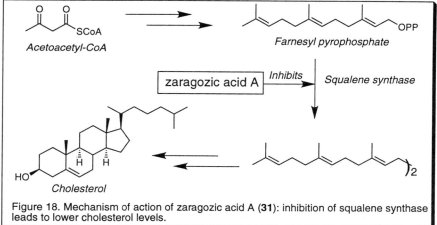

Acetoacetyl-CoA

Farnesyl pyrophosphate

zaragozic acid A → *Inhibits* → Squalene synthase

Cholesterol

Figure 18. Mechanism of action of zaragozic acid A (**31**): inhibition of squalene synthase leads to lower cholesterol levels.

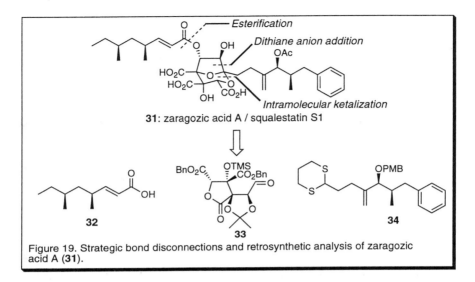

Figure 19. Strategic bond disconnections and retrosynthetic analysis of zaragozic acid A (**31**).

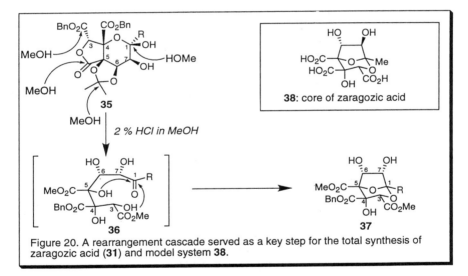

Figure 20. A rearrangement cascade served as a key step for the total synthesis of zaragozic acid (**31**) and model system **38**.

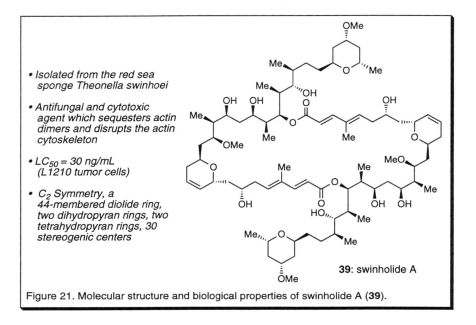

• Isolated from the red sea sponge Theonella swinhoei

• Antifungal and cytotoxic agent which sequesters actin dimers and disrupts the actin cytoskeleton

• LC_{50} = 30 ng/mL (L1210 tumor cells)

• C_2 Symmetry, a 44-membered diolide ring, two dihydropyran rings, two tetrahydropyran rings, 30 stereogenic centers

39: swinholide A

Figure 21. Molecular structure and biological properties of swinholide A (**39**).

In our group, the total synthesis[21-22] of swinholide A took a path chartered by retrosynthetically analyzing the molecule as shown in Figure 22 and defining structures **40** and **41** as key intermediates. While two sequential Yamaguchi-type couplings allowed the construction of the macrocyclic ring via C-O bond formations, the dithiane anion-cyclic sulfate coupling shown in Figure 23 served as the basis for the assembly of the required monomeric unit of swinholide A via C-C bond formation.

Despite the successes in total synthesis, however, the biological profile of swinholide and its potential in medicine remain largely unexplored.

Taxol™. Molecules from nature like taxol[23] are rare, but when they reveal themselves they provide much stimulation and excitement, and offer unique research opportunities. As a billion dollar drug for cancer therapy, taxol is a big success story and serves as an inspiring example of the rewards waiting to be harvested from natural products chemistry. Taxol's structure (**43**, Figure 24) is complex, highly congested, and adorned with reactive functionality to frustrate even the cleverest strategist for a long time. But what a wonderful playground that turned out to be: new synthetic technology and strategies are still emerging from the many programs initiated towards its total synthesis[24] and, indeed, several total syntheses[25] have already been reported. No doubt, several more will follow.

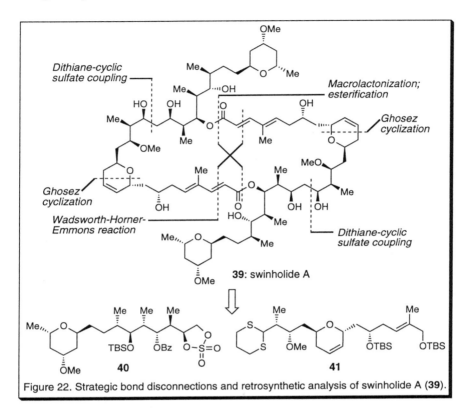

Figure 22. Strategic bond disconnections and retrosynthetic analysis of swinholide A (**39**).

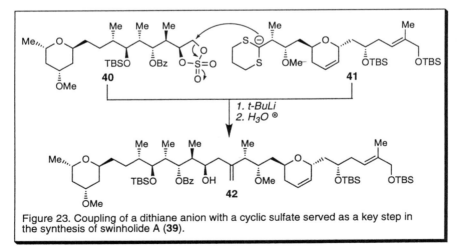

Figure 23. Coupling of a dithiane anion with a cyclic sulfate served as a key step in the synthesis of swinholide A (**39**).

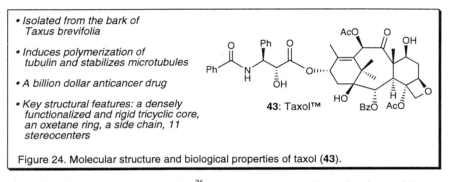

- Isolated from the bark of
  Taxus brevifolia

- Induces polymerization of
  tubulin and stabilizes microtubules

- A billion dollar anticancer drug

- Key structural features: a densely
  functionalized and rigid tricyclic core,
  an oxetane ring, a side chain, 11
  stereocenters

43: Taxol™

Figure 24. Molecular structure and biological properties of taxol (43).

Taxol's mechanism of action[26] is as intriguing as its molecular architecture. Discovered in 1979,[26] this mechanism involves induction of tubulin polymerization and microtubule stabilization (Figure 25). As a consequence, taxol severely hinders mitosis by blocking the division of the mitotic spindle (Figure 26). The net result is cell replication arrest and cell death. Taxol, originally isolated from the pacific yew tree,[23] is mimicked in its mechanism of action by the epothilones, isolated from soil bacteria, and the eluetherobins and sarcodictyins which come from soft corals. These last compounds will be discussed in the following section.

Striving for a convergent synthesis and inclusion of new methods in our synthetic strategy towards taxol, we utilized the bond disconnections and key building blocks shown in Figure 27. Thus, Shapiro and McMurry coupling reactions were employed in joining intermediates 45 and 46 and forming the 8-membered ring, while an intramolecular Diels-Alder reaction utilizing a temporary boron tethering device was applied in the construction of ring C (compound 50, Figure 28).

The knowledge gained during the campaign for the total synthesis facilitated the construction of a plethora of taxol analogs for biological investigations. Amongst the most interesting taxoids synthesized in our laboratories are the highly active agent 51 and the water-soluble and self-assembling cytotoxic compound 52 (Figure 29).[27]

With the total synthesis of taxol, the force of organic synthesis has been amply demonstrated, only to be challenged again by the latest and even more complex structures streaming from natural sources.

Epothilones. A new class of tubulin polymerization and microtubule stabilizing agents, the epothilones, appeared on the scene in the early 1990's. Isolated from myxobacteria,[28] found in a soil sample first collected from the banks of the Zimbase river in the Republic of South Africa, these substances combine novel molecular architecture, important biological activity, and interesting mechanism of action (Figure 30). Most significantly, epothilones A (53) and B (54) are, in addition to being more potent than taxol, active against multidrug-resistant (MDR) tumor cells, including taxol-resistant cell lines. The excitement about these new antitumor agents is reflected in the intense research activities they precipitated in chemistry and biology since their discovery. Being amongst those convinced of their potential and the opportunity they

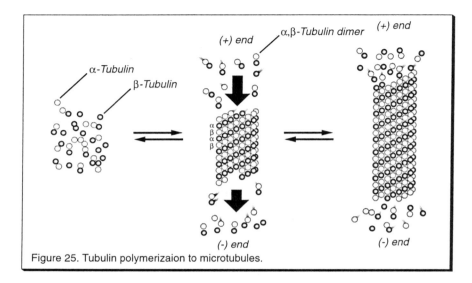

Figure 25. Tubulin polymerizaion to microtubules.

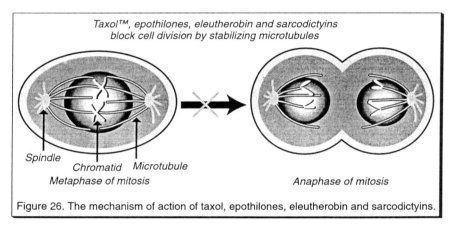

Figure 26. The mechanism of action of taxol, epothilones, eleutherobin and sarcodictyins.

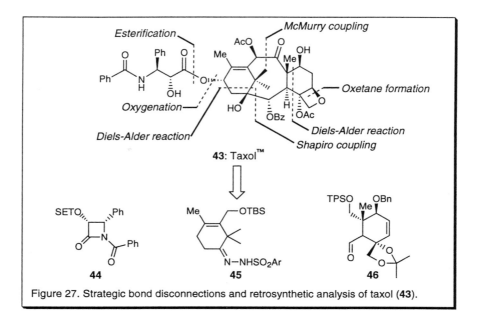

Figure 27. Strategic bond disconnections and retrosynthetic analysis of taxol (**43**).

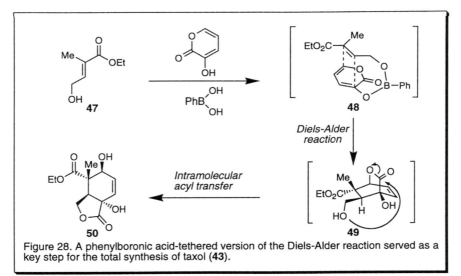

Figure 28. A phenylboronic acid-tethered version of the Diels-Alder reaction served as a key step for the total synthesis of taxol (**43**).

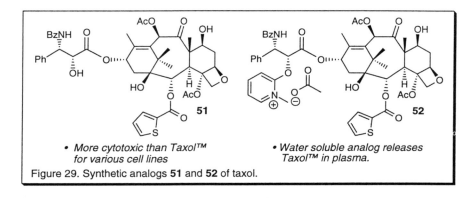

- More cytotoxic than Taxol™ for various cell lines

- Water soluble analog releases Taxol™ in plasma.

Figure 29. Synthetic analogs **51** and **52** of taxol.

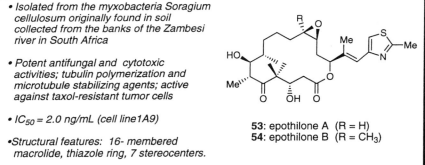

- *Isolated from the myxobacteria Soragium cellulosum originally found in soil collected from the banks of the Zambesi river in South Africa*

- *Potent antifungal and cytotoxic activities; tubulin polymerization and microtubule stabilizing agents; active against taxol-resistant tumor cells*

- $IC_{50} = 2.0$ ng/mL (cell line1A9)

- *Structural features: 16- membered macrolide, thiazole ring, 7 stereocenters.*

53: epothilone A (R = H)
54: epothilone B (R = CH_3)

Figure 30. Molecular structure and biological properties of epothilones A (**53**) and B (**54**).

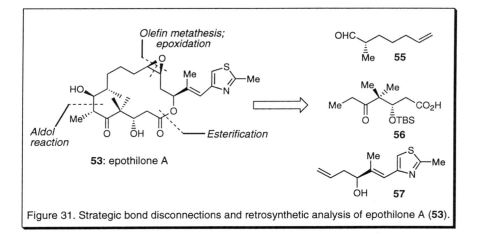

Figure 31. Strategic bond disconnections and retrosynthetic analysis of epothilone A (**53**).

offer in chemical biology, we initiated, in mid-1996, a program directed at their total synthesis.

From our several approaches to the epothilones, we present here perhaps the most aesthetically pleasing and productive strategy in terms of library generation (Figures 31 and 32). Our olefin metathesis strategy for the total synthesis of epothilones was successfully implemented both in solution[29] (Figure 31) and on solid phase[30] (Figure 32). The solid phase strategy, in particular, set a ground-breaking precedent for a new paradigm for complex natural product total synthesis and permitted the combinatorial synthesis of epothilone libraries[31] using radiofrequency encoded chemistry.[32] The chemical biology studies facilitated by the developed synthetic routes involved molecular design, chemical synthesis and biological evaluation of hundreds of epothilones. A number of these designed compounds exhibited promising biological profiles to warrant their further development as potential anticancer agents. Two of these structures (**61** and **62**) are shown in Figure 33.

The epothilone project exemplifies admirably the new trend in organic synthesis whereby the total synthesis of a target molecule is enriched by synthetic technology developments and chemical biology studies.[33] The epothilones have also been synthesized by two other groups.[34]

Eleutherobin and Sarcodictyin A. The excitement generated by taxol and the epothilones as antimitotic agents was intensified even further with the discovery of the tubulin-binding properties of sarcodictyin A (**63**)[35] and eleutherobin (**64**)[36] (Figure 34). Isolated from soft corals, these marine natural products are characterized by all the necessary features that make an exciting chemical biology program: novel molecular frameworks, important biological activity and intriguing mechanism of action. Their potential in cancer chemotherapy was not missed, neither by the pharmaceutical industry nor by us.

Focusing on a strategy derived from the bond disconnections and retrosynthetic analysis shown in Figure 35, and starting from the readily available (+)-carvone, we were soon able to synthesize both sarcodictyin A (**63**)[37] and eleutherobin (**64**).[38] The novel ring closure which occurred upon selective hydrogenation of the acetylenic bond in **68** served admirably in casting the tricyclic skeleton of eleutherobin (and sarcodictyin) as shown in Figure 36.

This ongoing program is currently shaping into another exemplary project in chemical biology with exciting aspects of molecular design, combinatorial library generation, and biological evaluation. Tubulin-binding and tumor cytotoxicity assays have already identified library members **71** and **72** (Figure 37) as highly active lead compounds.

Brevetoxin B. The brevetoxins are a class of highly potent neurotoxins isolated from the marine organism dinoflagellate *Gymnodinium breve* which is associated with the "red tide" phenomena.[39] These destructive and often visible blooms of unicellular algae (phytoplankton) occur periodically along certain areas of the world's coastal waters and cause catastrophic killings of fish and other marine life as well as affecting human health through poisoned shellfish. The "red tides" are not a new phenomenon; their history is recorded in the fossils, and in the Bible, as believed by some. However, the modern world is facing an alarming spate of reported incidents, possibly

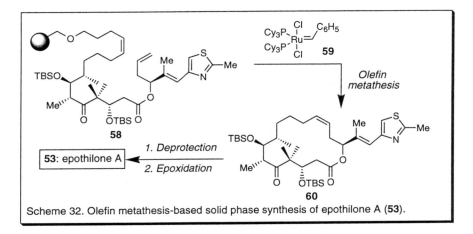

Scheme 32. Olefin metathesis-based solid phase synthesis of epothilone A (**53**).

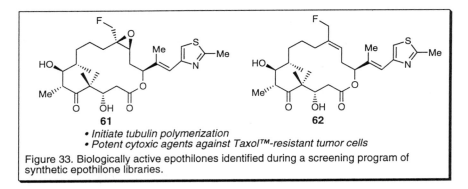

* Initiate tubulin polymerization
* Potent cytoxic agents against Taxol™-resistant tumor cells

Figure 33. Biologically active epothilones identified during a screening program of synthetic epothilone libraries.

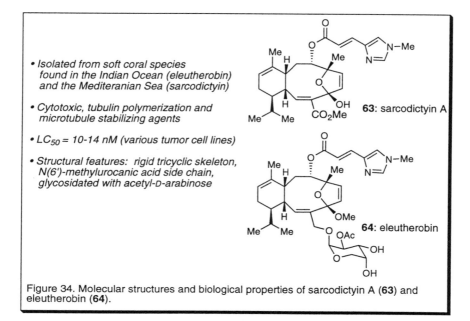

- *Isolated from soft coral species found in the Indian Ocean (eleutherobin) and the Mediteranian Sea (sarcodictyin)*

- *Cytotoxic, tubulin polymerization and microtubule stabilizing agents*

- *LC$_{50}$ = 10-14 nM (various tumor cell lines)*

- *Structural features: rigid tricyclic skeleton, N(6')-methylurocanic acid side chain, glycosidated with acetyl-D-arabinose*

63: sarcodictyin A

64: eleutherobin

Figure 34. Molecular structures and biological properties of sarcodictyin A (**63**) and eleutherobin (**64**).

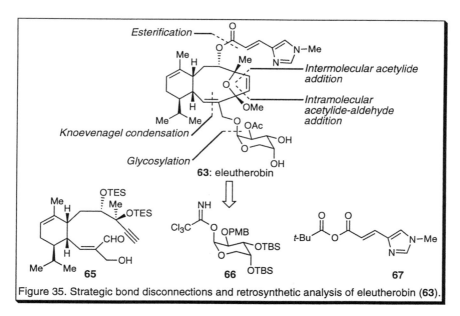

Figure 35. Strategic bond disconnections and retrosynthetic analysis of eleutherobin (**63**).

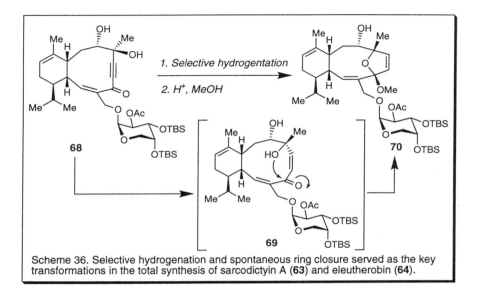

Scheme 36. Selective hydrogenation and spontaneous ring closure served as the key transformations in the total synthesis of sarcodictyin A (**63**) and eleutherobin (**64**).

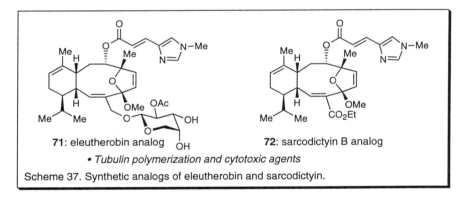

71: eleutherobin analog

72: sarcodictyin B analog

• *Tubulin polymerization and cytotoxic agents*

Scheme 37. Synthetic analogs of eleutherobin and sarcodictyin.

due to an increase in nutrient-rich waters caused by human pollution and world-wide shipping practices. Brevetoxin B (**73**, Figure 38) was the first member of the brevetoxin family to be discovered, and its structure was fully elucidated by spectroscopic and X-ray crystallographic techniques.[40] Structurally, brevetoxin B possesses an unprecedented molecular framework composed of rings, *trans*-fused in a ladder-like manner. Its complexity and architectural beauty stems from weaving a single carbon chain into a series of rings using oxygen bridges with striking regularity (Figure 38) forming a polycyclic framework of common and medium size rings, each containing an oxygen atom. Brevetoxin A, a similarly behaving neurotoxin secreted from the same marine organism, will be discussed in the subsequent section.

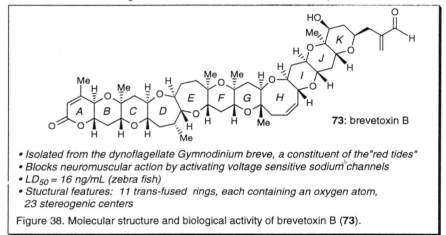

• *Isolated from the dynoflagellate Gymnodinium breve, a constituent of the"red tides"*
• *Blocks neuromuscular action by activating voltage sensitive sodium channels*
• *LD_{50} = 16 ng/mL (zebra fish)*
• *Stuctural features: 11 trans-fused rings, each containing an oxygen atom, 23 stereogenic centers*

Figure 38. Molecular structure and biological activity of brevetoxin B (**73**).

The biological mode of action of the brevetoxins involves binding and activation of membrane-bound sodium channels (Figure 39) which serve vital roles in neurological communication and which also serve as the targets for many other known toxins (e.g. puffer fish toxin, snake venom). By the nature of their molecular structures and hydrophobic properties, the brevetoxins are able to penetrate the cell membrane and selectively bind to these ion channels, thus altering the protein's conformation and functions. Ultimately, this binding precipitates an influx of sodium ions resulting in several assailing neurological and physiological effects such as disorientation and muscle cramps.

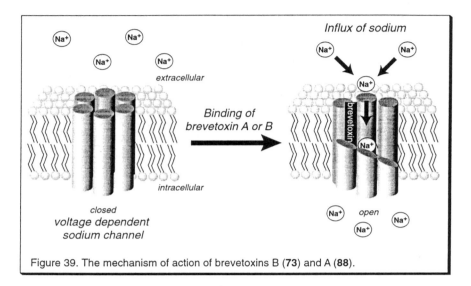

Figure 39. The mechanism of action of brevetoxins B (**73**) and A (**88**).

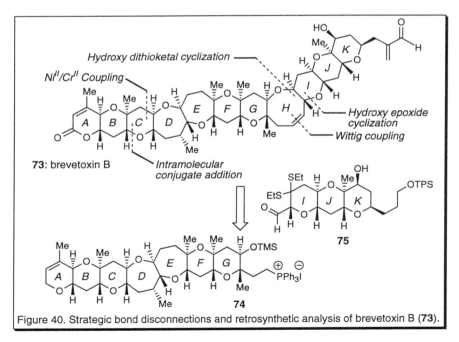

Figure 40. Strategic bond disconnections and retrosynthetic analysis of brevetoxin B (**73**).

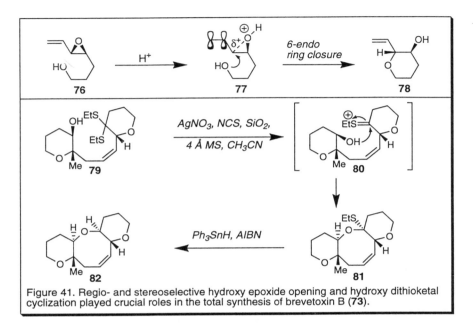

Figure 41. Regio- and stereoselective hydroxy epoxide opening and hydroxy dithioketal cyclization played crucial roles in the total synthesis of brevetoxin B (**73**).

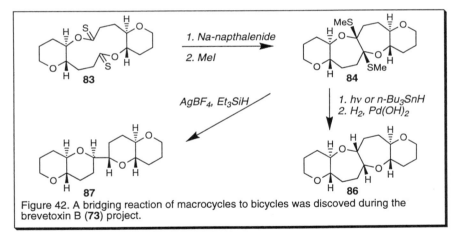

Figure 42. A bridging reaction of macrocycles to bicycles was discoved during the brevetoxin B (**73**) project.

Due to the highly complex nature of the brevetoxin B molecule (73), the synthetic strategy had to be redesigned several times and numerous new reactions had to be invented and developed before final success.[41] Figure 40 shows the strategic bond disconnections and key intermediates utilized in the final plan that culminated in a convergent total synthesis of brevetoxin B (73). During the 12-year synthetic odyssey[42] that ultimately led to the targeted molecule, many new reactions were discovered and perfected. Amongst the most powerful and practical methodologies unearthed during this program were the 6-*endo* activated hydroxy epoxide cyclizations for the construction of tetrahydropyran systems,[43] the hydroxy dithioketal cyclization to form oxecane systems,[44] and the bridging of macrocycles to bicycles[45] (Figures 41 and 42). In addition, a series of interesting oxygen- and sulfur-containing molecular structures of both chemical and theoretical interest emerged from this program.[45]

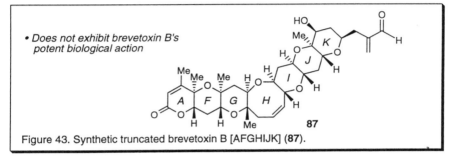

Figure 43. Synthetic truncated brevetoxin B [AFGHIJK] (87).

With regards to chemical biology studies, the truncated brevetoxin B analog **87** (Figure 43) was synthesized[46] and studied[47] for its interaction with the sodium channel. It was found to lack brevetoxin B's biological actions, suggesting that the entire length of this neurotoxin is rather important for binding and triggering sodium influx.

The wealth of information gathered during the brevetoxin B campaign was enormous, worthwhile and rewarding, and is surpassed only by a few synthetic projects. One such project was perhaps that directed at the total synthesis of brevetoxin A, brevetoxin B's more beautiful sister molecule discussed below.

Brevetoxin A. The structural elucidation[48] of the most potent marine neurotoxin secreted by *Gymnodinium breve* (the producer of brevetoxin B), brevetoxin A (**88**, Figure 44) presented yet another "Mount Everest" and an opportunity to discover and develop new chemistry. Even though this target resembles, somewhat, brevetoxin B, the inclusion of a 9-membered ring, in addition to the three 8-membered, one 7-membered, four 6-membered and one 5-membered rings, complicated the task of its total synthesis[49] far beyond the construction of brevetoxin B. New synthetic chemistry and novel synthetic strategies had to be developed for the new problem at hand. The final strategy used was charted along a path revealed by the bond disconnections and retrosynthetic analysis shown in Figure 45. Two of the most satisfying developments

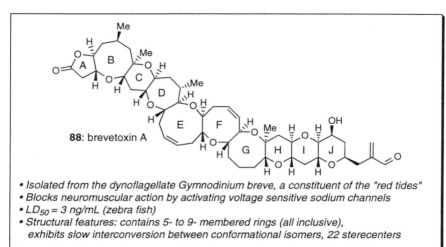

- *Isolated from the dynoflagellate Gymnodinium breve, a constituent of the "red tides"*
- *Blocks neuromuscular action by activating voltage sensitive sodium channels*
- *LD$_{50}$ = 3 ng/mL (zebra fish)*
- *Structural features: contains 5- to 9- membered rings (all inclusive), exhibits slow interconversion between conformational isomers, 22 sterecenters*

Figure 44. Molecular structure and biological properties of brevetoxin A (**88**).

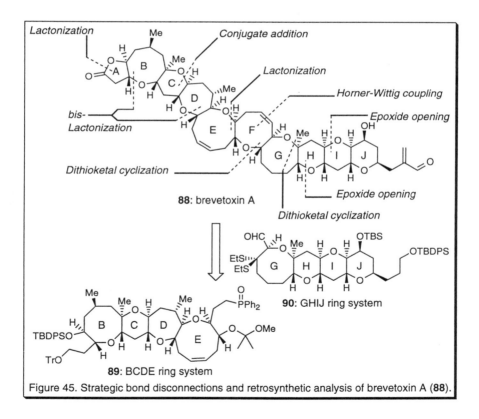

Figure 45. Strategic bond disconnections and retrosynthetic analysis of brevetoxin A (**88**).

in terms of synthetic technology and strategies were the palladium-catalyzed C-C bond forming reaction involving cyclic ketene acetal phosphates and vinyltin reagents,[50] specifically devised to solve the medium sized rings of brevetoxin A, and the singlet oxygen functionalization of the 9-membered ring[51] as summarized in Figure 46.

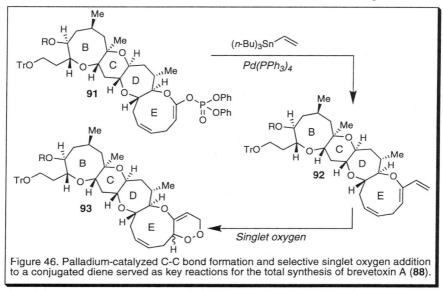

Figure 46. Palladium-catalyzed C-C bond formation and selective singlet oxygen addition to a conjugated diene served as key reactions for the total synthesis of brevetoxin A (**88**).

The mastery of a synthetic route to brevetoxin A left us with the tools and experience to synthesize a series of designed brevetoxins for chemical biology studies. Indeed, future progress in sodium channel research and neurobiology, in general, may depend to a large extent on organic synthesis.

Perspectives

In this article, we discussed the genesis and evolution of organic synthesis to its present day prominence within the sciences in general and within chemistry, in particular. It is very clear that this discipline has been, and will continue to be, highly enabling to many endeavors, and particularly to biology and medicine. The issues before us are how to advance the science of organic synthesis beyond its present boundaries, and where and how to apply it most effectively in the future for the benefit of humankind. From our present standpoint, a number of statements can be made with certainty: (1) the art and science of organic synthesis as potent as it may seem, it is far from ideal or even mature, if we compare it to biosynthesis; (2) its potential for creating new substances for untold applications is enormous, and is only limited by human imagination and humanpower. Organic synthesis will be enriched by discoveries and inventions, which in turn will accelerate and make possible major breakthroughs in biology and medicine, as well as in other areas of science and technology. But, how are these discoveries and inventions going to happen? How can we optimize our chances for success and in what fields we should focus our efforts? What new directions should we initiate in organic synthesis and within chemistry in general? Most of us are pondering such questions, a few are even pontificating! At the risk of being placed in the latter category, an inclusion which might not be necessarily honorable, we will attempt to speculate on these issues.

The discoveries of the future in organic synthesis will come from rationally designed research programs directed at solving currently perceived and recognized problems, such as the synthesis of enantiomerically pure and high value intermediates, or the total synthesis of complex and scarce natural substances for biological investigations. But looking into the future is like walking on the surface of a sphere where the horizon takes us only so far ahead. As we move forward and stand on new ground, we can extend our vision even further and move into new areas of opportunity which we cannot see or imagine today. This realization should somewhat ease our anxieties about making precise long term predictions, even though vision should always play an important role in our thinking. We should always remember that those who are willing and courageous to attempt the seemingly impossible and to maneuver themselves away from rigid and obsolete plans and into newly emerging opportunities are rewarded handsomely. As in the past, a third avenue to discovery in the future will be serendipity. Such unpredictable discoveries (e.g. Wittig reaction, Teflon, penicillin) had an enormous impact on chemistry and society in the past and will, no doubt, occur in and shape our future. No doubt, the present frontiers of new synthetic technology, catalysis, material science, supramolecular assemblies, total synthesis, chemical biology and medicinal chemistry will be continuously pushed forward through vigorous research activities.

Our own vision is driven by nature's beauty and wealth in terms of molecular diversity, biological activity and medicine. Indeed, we envision that the structures of penicillin, calicheamicin, Taxol™, brevetoxin and eleutherobin will soon be joined by many others of even more stunning molecular beauty and biological activity. Surely, nature has not revealed to us but a small fraction of its molecular secrets and as we pursue their discovery, we will learn from them and we will be challenged by them. Imaginative molecular designs and combinatorial chemistry will also play a central role in our research programs as we attempt to facilitate and enable further the drug discovery process and improve the capabilities of medicinal chemistry in its quest for new therapies. Thus, the synthetic chemist will continue to be confronted by intriguing challenges and opportunities to create and invent new science in the fields of synthetic technology, total synthesis and chemical biology. Properly faced and exploited, such opportunities will deliver rewards of untold value to society through chemistry, biology and medicine.

Finally, it is incumbent upon ourselves to ensure the harmonious and synergistic interplay between government, academia and industry in order to ensure strong basic research and optimize its impact on the nation's economy and welfare. In doing so we must always keep in mind the preservation of our environment and the maintenance of nature's harmony. This philosophy served us well in the past, and to our children we owe its continuance.

Acknowledgements. It is with great pride and pleasure that we wish to thank our collaborators whose names appear in the references and whose contributions made the described work possible and enjoyable. We gratefully acknowledge the National Institutes of Health USA, Merck Sharp & Dohme, DuPont Merck, Schering-Plough,

Pfizer, Hoffmann-La Roche, Glaxo, Rhone-Poulenc Rorer, Aldrich Chemical Co., Amgen, Novartis, CaP CURE and the Skaggs Institute for Chemical Biology for supporting our research programs.

Literature Cited

1. Nicolaou, K. C.; Sorenson, E. *Classics in Total Synthesis;* VCH: Weinheim, Germany, 1996, pp 1-789.
2. Breslow, R. *Chemistry: Today and Tomorrow: The Central, Useful and Creative Science;* American Chemical Society: Washington D.C., 1996, pp 1-144.
3. Lee, M. D.; Dunne, T. S.; Siegel, M. M.; Chang, C. C.; Morton, G. O.; Borders, D. B. *J. Am. Chem. Soc.* **1987**, *109*, pp 3464-3466; Lee, M. D.; Dunne, T. S.; Chang, C. C.; Ellestad, G. A.; Siegel, M. M.; Morton, G. O.; McGahren, W. J.; Borders, D. B. *J. Am. Chem. Soc.* **1987**, *109*, pp 3466-3468.
4. Nicolaou, K. C.; Hummel, C. W.; Pitsinos, E. N.; Nakada, M.; Smith, A. L.; Shibayama, K.; Saimoto, H. *J. Am. Chem. Soc.* **1992**, *114*, pp 10082-4.
5. Nicolaou, K. C.; Dai, W.-M. *Angew. Chem. Int. Ed. Engl.* **1991**, *30*, pp 1387-1416.
6. Nicolaou, K. C.; Smith, A. L.; Yue, E. W. *Proc. Natl. Acad. Sci., USA* **1993**, *90*, pp 5881-8.
7. Nicolaou, K. C. *Angew. Chem. Int. Ed. Engl.* **1993**, *32*, pp 1377-85.
8. Rosen, M. K.; Schreiber, S. L. *Angew. Chem., Int. Ed. Engl.* **1992**, *31*, pp 384-400.
9. Swindells, D. C. N.; White, P. S.; Findlay, J. A. *Can. J. Chem.* **1978**, *56*, pp 2491-2492. Findlay, J. A.; Radics, L. *Can. J. Chem.* **1980**, *58*, pp 579-590.
10. Nicolaou, K. C.; Chakraborty, T. K.; Piscopio, A. D.; Minowa, N.; Bertinato, P. *J. Am. Chem. Soc.* **1993**, *115*, pp 4419-20.
11. Romo, D.; Meyer, S. D.; Johnson, D. D.; Schreiber, S. L. *J. Am. Chem. Soc.* **1993**, *115*, pp 7906-7907. Hayward, C. M.; Yohannes, D.; Danishefsky, S. J. *J. Am. Chem. Soc.* **1993**, *115*, pp 9345-9346. Smith III, A. B.; Condon, S. M.; McCauley, J. A.; Leaser Jr., J. L.; Leahy, J. W.; Maleczka Jr., R. E. *J. Am. Chem. Soc.* **1995**, *117*, pp 5407-5408.
12. Chakraborty, T. K.; Weber, H. P.; Nicolaou, K. C. *Chem. Biol.* **1995**, *2*, pp 157-61.
13. Kulanthaivel, P.; Hallock, Y. F.; Boros, C.; Hamilton, S. M.; Janzen, W. P.; Ballas, L. M.; Loomis, C. R.; Jiang, J. B.; Katz, B.; Steiner, J. R.; Clardy, J. *J. Am. Chem. Soc.* **1993**, *115*, pp 6452-6453.
14. Nishizuka, Y. *Science* **1992**, *258*, pp 607-614.
15. Nicolaou, K. C.; Bunnage, M. E.; Koide, K. *J. Am. Chem. Soc.* **1994**, *116*, pp 8402-3.
16. Koide, K.; Bunnage, M. E.; Paloma, L. G.; Kanter, J. R.; Taylor, S. S.; Brunton, L. L.; Nicolaou, K. C. *Chem. Biol.* **1995**, *2*, pp 601-8.
17. Bergstrom, J. D.; Kurtz, M. M.; Rew, D. J.; Amend, A. M.; Karkas, J. D.; Bostedor, R. G.; Bansal, V. S.; Dufresne, C.; VanMiddlesworth, F. L.; Hensens, O. D.; Liesch, J. M.; Zink, D. L.; Wilson, K. E.; Onishi, J.; Milligan, J. A.; Bills, G.; Kaplan, L.; Nallin, M.; Omstead, M. N.; Jenkins, R. G.; Huang, L.; Meinz, M. S.; Quinn, L.; Burg, R. W.; Kong, Y. L.; Mochales, S.; Mojena, M.; Martin, I.; Pelaez, F.; Diez, M. T.; Alberts, A. W. *Proc. Natl. Acad. Sci., U.S.A.* **1993**, *90*, pp 80-84. Sidebottom, P. J.; Highcock, R. M.; Lane, S. J.; Procopiou, P. A.; Watson, N. S. *J. Antibiot.* **1992**, *45*, pp 648-58.

18. Nadin, A.; Nicolaou, K. C. *Angew. Chem. Int. Ed. Engl.* **1996**, *35*, pp 1622-56. Procopiou, P. A.; Cox, B.; Kirk, B. E.; Lester, M. G.; McCarthy, A. D.; Sareen, M.; Sharratt, P. J.; Snowden, M. A.; Spooner, S. J.; Watson, N. S.; Widdowson, J. *J. Med. Chem.* **1996**, *39*, pp 1413-1422.

19. Nicolaou, K. C.; Yue, E. W.; Naniwa, Y.; De Riccardis, F.; Nadin, A.; Leresche, J. E.; La Greca, S.; Yang, Z. *Angew. Chem. Int. Ed. Engl.* **1994**, *33*, 2184-2187. Nicolaou, K. C.; Nadin, A.; Leresche, J. E.; La Greca, S.; Tsuri, T.; Yue, E. W.; Yang, Z. *Angew. Chem. Int. Ed. Engl.* **1994**, *33*, pp 2187-2190. Nicolaou, K. C.; Nadin, A.; Leresche, J. E.; Yue, E. W.; La Greca, S. *Angew. Chem. Int. Ed. Engl.* **1994**, *33*, pp 2190-91. Carreira, E. M.; Dubois, J. *J. Am. Chem. Soc,* **1994**, *116*, pp 10825-10826. Evans, D. A.; Barrow, J. C.; Leighton, J. L.; Robichaud, A. J.; Sefkow, M. J. *J. Am. Chem. Soc.* **1994**, *116*, pp 12111-12112.

20. Carmely, S.; Kashman, Y. *Tetrahedron Lett.* **1985**, *26*, pp 511-514. Kobayashi, M.; Tanaka, J.; Katori, T.; Matsuura, M.; Kitagawa, I. *Tetrahedron Lett.* **1989**, *30*, pp 2963-2966. Kobayashi, M.; Tanaka, J.; Katori, T.; Matsuura, M.; Yamashita, M.; Kitagawa, I. *Chem. Pharm. Bull.* **1990**, *38*, pp 2409-2418. Kitagawa, I.; Kobayashi, M.; Katori, T.; Yamashita, M.; Tanaka, J.; Doi, M.; Ishida, T. *J. Am. Chem. Soc.* **1990**, *112*, pp 3710-3712. Doi, M.; Ishida, T.; Kobayashi, M.; Kitagawa, I. *J. Org. Chem.* **1991**, *56*, pp 3629-3632.

21. Nicolaou, K. C.; Ajito, K.; Patron, A. P.; Khatuya, H.; Richter, P. K.; Bertinato, P. *J. Am. Chem. Soc.* **1996**, *118*, pp 3059-60.

22. Paterson, I.; Yeung, K.; Ward, R. A.; Cumming, J. G.; Smith, J. D. *J. Am. Chem. Soc.* **1994**, *116*, pp 9391-9392. Paterson, I.; Yeung, K.; Ward, R. A.; Smith, J. D.; Cumming, J. G.; Lamboley, S. *Tetrahedron* **1995**, *51*, pp 9467-9486.

23. Wani, M. C.; Taylor, H. L.; Wall, M. E.; Coggen, P.; McPhail, A. T. *J. Am. Chem. Soc.* **1971**, *93*, pp 2325-2327.

24. Nicolaou, K. C.; Dai, W.-M.; Guy, R. K. *Angew. Chem. Int. Ed. Engl.* **1994**, *33*, pp 15-44.

25. Nicolaou, K. C.; Yang, Z.; Liu, J. J.; Ueno, H.; Nantermet, P. G.; Guy, R. K.; Claiborne, C. F.; Renaud, J.; Couladouros, E. A.; Paulvannan, K.; Sorensen, E. J. *Nature* **1994**, *367*, pp 630-634. Holton, R. A.; Somoza, C.; Kim, H. B.; Liang, F.; Biediger, R. J.; Boatman, P. D.; Shindo, M.; Smith, C. C.; Kim, S.; Suzuki, Y.; Tao, C.; Vu, P.; Tang, S.; Zhang, P.; Murthi, K. K.; Gentile, L. N.; Liu, J. H. *J. Am. Chem. Soc.* **1994**, *116*, pp 1597-1598. Holton, R. A.; Kim, H. B.; Somoza, C.; Liang, F.; Biediger, R. J.; Boatman, P. D.; Shindo, M.; Smith, C. C.; Kim, S.; Nadizadeh, H.; Suzuki, Y.; Tao, C.; Vu, P.; Tang, S.; Zhang, P.; Murthi, K. K.; Gentile, L. N.; Liu, J. H. *J. Am. Chem. Soc.* **1994**, *116*, pp 1599-1600. Danishefsky, S. J.; Masters, J. J.; Young, W. B.; Link, J. T.; Snyder, L. B.; Magee, T. V.; Jung, D. K.; Isaacs, R. C.; Bornmann, W. G.; Alaimo, C. A.; Coburn, C. A.; Di Grandi, M. J. *J. Am. Chem. Soc.* **1996**, *118*, pp 2843-2859. Wender, P. A.; Badham, N. F.; Conway, S. P.; Floreancig, P. E.; Glass, T. E.; Gränicher, C.; Houze, J. B.; Jänichen, J.; Lee, D.; Marquess, D. G.; McGrane, P. L.; Meng, W.; Mucciaro, T. P.; Mühlebach, M.; Natchus, M. G.; Paulsen, H.; Rawlins, D. B.; Satkofsky, J.; Shuker, A. J.; Sutton, J. C.; Taylor, R. E.; Tomooka, K. *J. Am. Chem. Soc.* **1997**, *119*, pp 2755-2756. Wender, P. A.; Badham, N. F.; Conway, S. P.; Floreancig, P. E.; Glass, T. E.; Houze, J. B.; Krauss, N. E.; Lee, D.; Marquess, D. G.; McGrane, P. L.; Meng, W.; Natchus, M. G.; Shuker, A. J.; Sutton, J. C.; Taylor, R. E. *J. Am. Chem. Soc.* **1997**, *119*, pp 2757-2758. Mukaiyama, T.; Shina,

I.; Iwadare, H.; Sakoh, H.; Tani, Y.-I.; Hasegawa, M.; Saitoh, K. *Proc. Jpn. Acad., Ser. B*, **1997**, *73B*, pp 95-100.

26. Horwitz, S. B.; Fant, J.; Schiff, P. B. *Nature* **1979**, *277*, pp 665-667.

27. Nicolaou, K. C.; Guy, R. K.; Pitsinos, E. N.; Wrasidlo, W. *Angew. Chem. Int. Ed. Engl.* **1994**, *33*, pp 1583-1587. Nicolaou, K. C.; Couladouros, E. A.; Nantermet, P. G.; Renaud, J.; Guy, R. K.; Wrasidlo, W. *Angew. Chem. Int. Ed. Engl.* **1994**, *33*, pp 1581-1583.

28. Höfle, G.; Bedorf, N.; Gerth, K.; Reichenbach, H. (GHF), DE-4138042, **1993** [Chem. Abstr., 120, 52841 (1993)]; Gerth, K.; Bedorf, N.; Höfle, G.; Irschik, H.; Reichenbach, H. *J. Antibiot.* **1996**, *49*, pp 560-563. Höfle, G.; Bedorf, N.; Steinmetz, H.; Schomburg, D.; Gerth, K.; Reichenbach, H. *Angew. Chem. Int. Ed. Engl.* **1996**, *35*, pp 1567-1569.

29. Nicolaou, K. C.; Ninkovic, S., Sarabia, F.; Vourloumis, D.; He, Y.; Vallberg, H.; Finlay, M. R. V.; Yang, Z. *J. Am. Chem. Soc.* **1997**, *119*, pp 7974-7991.

30. Nicolaou, K. C.; Winssinger, N.; Pastor, J.; Ninkovic, S.; Sarabia, F.; He, Y.; Vourloumis, D.; Yang, Z.; Li, T.; Giannakakou, P.; Hamel, E. *Nature* **1997**, *387*, pp 268-272.

31. Nicolaou, K. C.; Vourloumis, D.; Li, T.; Pastor, J.; Winssinger, N.; He, Y.; Ninkovic, S.; Sarabia, F.; Vallberg, H.; Roschangar, F.; King, N. P.; Finlay, M. R. V.; Giannakakou, P.; Verdier-Pinard, P.; Hamel, E. *Angew. Chem. Int. Ed. Engl.* **1997**, *36*, pp 2097-2103.

32. Nicolaou, K. C.; Xiao, X.-Y.; Parandoosh, Z.; Senyei, A.; Nova, M. P. *Angew. Chem., Int. Ed. Engl.* **1995**, *34*, pp 2289-2291. Moran, E. J.; Sarshar, S.; Cargill, J. F.; Shahbaz, M. M.; Lio, A.; Mjalli, A. M. M.; Armstrong, R. W. *J. Am. Chem. Soc.* **1995**, *117*, pp 10787-10788.

33. Nicolaou, K. C.; Roschangar, F.; Vourloumis, D. *Angew. Chem. Int. Ed. Engl.* in press.

34. Meng, D.; Bertinato, P.; Balog, A.; Su, D.-S.; Kamenecka, T.; Sorensen, E. J.; Danishefsky, S. J. *J. Am. Chem. Soc.* **1997**, *119*, pp 10073-10092; Schinzer, D.; Limberg, A.; Böhm, O. M.; Cordes, M. *Angew. Chem. Int. Ed. Eng.* **1997**, *36*, pp 2520-2524.

35. Fenical, W.-H.; Hensen, P. R.; Lindel, T. U.S. Patent No. 5,473,057, Dec. 5, 1995. Lindel, T.; Jensen, P. R.; Fenical, W.; Long, B. H.; Casazza, A. M.; Carboni, J.; Fairchild, C. R. *J. Am. Chem. Soc.* **1997**, *119*, pp 8744-8745.

36. D'Ambrosio, M.; Guerriero, A.; Pietra, F. *Helv. Chim. Acta* **1987**, *70*, pp 2019-2027. D'Ambrosio, M.; Guerriero, A.; Pietra, F. *Helv. Chim. Acta* **1988**, *71*, pp 964-976.

37. Nicolaou, K. C.; Xu, J.-Y.; Kim, S.; Ohshima, T.; Hosokawa, S.; Pfefferkorn, J. *J. Am. Chem. Soc.* **1997**, *119*, pp 11353-11354.

38. Nicolaou, K. C.; van Delft, F.; Ohshima, T.; Vourloumis, D.; Xu, J.; Hosokawa, S.; Pfefferkorn, J.; Kim, S.; Li, T. *Angew. Chem. Int. Ed. Eng.* **1997**, *36*, pp 2520-2524.

39. Anderson, D. M. *Scientific American* **1994**, *271*, pp 62-68.

40. Lin, Y.-Y.; Risk, M.; Ray, S. M.; Van Engen, D.; Clardy, J.; Golik, J.; James, J. C.; Nakanishi, K. *J. Am. Chem. Soc.* **1981**, *103*, pp 6773-6775.

41. Nicolaou, K. C.; Theodorakis, E. A.; Rutjes, F. P. J. T.; Theodorakis, E. A.; Tiebes, J.; Sato, M.; Untersteller, E. *J. Am. Chem. Soc.* **1995**, *117*, pp 1171-1172.

Nicolaou, K. C.; Rutjes, F. P. J. T.; Theodorakis, E. A.; Tiebes, J.; Sato, M.; Untersteller, E. *J. Am. Chem. Soc.* **1995**, *117*, pp 1173-1174.

42. Nicolaou, K. C. *Angew. Chem. Int. Ed. Engl.* **1996**, *35*, pp 589-607.

43. Nicolaou, K. C.; Prasad, C. V. C.; Somers, P. K.; Hwang, C.-K. *J. Am. Chem. Soc.* **1989**, *111*, pp 5335-40.

44. Nicolaou, K. C.; Prasad, C. V. C.; Hwang, C.-K.; Duggan, M. E.; Veale, C. A. *J. Am. Chem. Soc.* **1989**, *111*, pp 5321-5330.

45. Nicolaou, K. C.; Hwang, C.-K.; Marron, B. E.; DeFrees, S..A.; Couladouros, E. A.; Abe, Y.; Carrol, P. J.; Snyder, J. *J. Am. Chem. Soc.* **1990**, *112*, pp 3040-3054; Nicolaou, K. C.; DeFrees, S. A.; Hwang, C. K.; Stylianides, N.; Carroll, P. J.; Snyder, J. P. *J. Am. Chem. Soc.*, **1990**, *112*, pp 3029-3039.

46. Nicolaou, K. C.; Tiebes, J.; Theodorakis, E. A.; Rutjes, F. P. J. T.; Sato, M.; Untersteller, E. *J. Am. Chem. Soc.* **1994**, *116*, pp 9371-9372.

47. Gawley, R. E.; Rein, K. S.; Jeglitsch, G.; Adams, D. J.; Theodorakis, E. A.; Tiebes, J.; Nicolaou, K. C.; Baden, D. G. *Chem. Biol.* **1995**, *2*, pp 533-541.

48. Shimizu, Y.; Chou, H.-N.; Bando, H.; Van Duyne, G.; Clardy, J. *J. Am. Chem. Soc.* **1986**, *108*, pp 514-515.

49. Nicolaou, K. C.; Yang, Z.; Shi, G.-Q.; Gunzner, J. L.; Agrios, K.A.; Gärtner, P. *Nature*, in press.

50. Nicolaou, K. C.; Shi, G.-Q.; Gunzner, J. L.; Gärtner, P.; Yang, Z. *J. Am. Chem. Soc.* **1997**, *119*, pp 5467-5468.

51. Nicolaou, K. C.; Yang, Z.; Ouellette, M.; Shi, G.-Q.; Gärtner, P.; Gunzner, J. L.; Agrios, C.; Huber, R.; Chadha, R.; Huang, D. H. *J. Am. Chem. Soc.* **1997**, *119*, pp 8105-8106.

7

The Pivotal Role of Chemistry in the Development of New Drug Discovery Paradigms in the Pharmaceutical Industry

Alan J. Main

Senior Vice President for Research, Novartis Pharmaceuticals Corp.
556 Morris Avenue, Summit, NJ 07901

Two new drug discovery paradigms in the pharmaceutical industry are presented, namely ''standardize the product'' where a single chemical class of compounds (antisense oligonucleotides), is being developed as potential universal therapeutic agents; and "standardize the process" where the drug discovery process itself is being highly optimized to efficiently deliver new therapies. The pivotal role that the discipline of chemistry is playing in the realization of these two alternative strategies is highlighted.

Introduction

Over the past few years, the pharmaceutical industry has been under pressure from a number of different directions as both the external scientific and regulatory environment have dramatically changed. On the regulatory front, government pressure on the pricing of drugs has increased worldwide and legislation has made it easier for generic equivalents to be marketed which rapidly erode the sales of products when they come off patent. In the market place the emergence of healthcare organizations who focus on the reduction of pharmaceutical costs have also contributed to reducing the potential profitability of new and current products. Simultaneously, on the scientific front, new competition such as "lean and mean" biotechnology companies have entered the arena. The areas of unmet medical need have moved from diseases that can be treated acutely (e.g. antibiotic treatment of bacterial infections) or symptomatically (e.g. hypertension) to much more complex and chronic diseases such as cancer, Alzheimer's, and osteoarthritis etc. The above factors have necessitated the application of the increasingly sophisticated (and expensive!) techniques of genomics, combinatorial chemistry, protein structural determination (X-Ray and NMR), and molecular modelling etc.   In addition, the growing understanding of fundamental disease mechanisms coupled with diagnostic methods are beginning to sub-segment diseases into more homogeneous but smaller patient populations, again reducing the potential revenue for each marketed drug.

The market place is therefore demanding better drugs, developed faster, at a lower price, in smaller overall markets.   Truly a formidable challenge to the pharmaceutical industry!   This demand for increased quality of products at lower prices is not unique to the pharmaceutical industry.   Virtually all major industries have been faced with this challenge.   To respond to this challenge, the pharmaceutical industry has been developing new paradigms in drug discovery, with the discipline of chemistry playing a key role.   In analogy to the manufacturing industry, two fundamental strategies are emerging, namely; "standardize the product" and "standardize the R and D process".

Standardize the Product

The current products of the pharmaceutical industry i.e. drugs, are all unique and vary widely from each other in their chemical composition.   Some are complex proteins such as human growth hormone, others are polysaccharides such as heparin, small peptides like oxytocin, natural products like erythromycin, simple organic molecules e.g. N-hydroxyurea or complex molecules synthesized in more than 20-30 chemical steps like HIV protease inhibitors.   This wide variety of structural classes of drugs requires a multiplicity of systems to handle these diverse chemical entities.   Each class requires highly specialized expertise to discover or invent these new agents, whether it be molecular biology to discovery new biopharmaceuticals, microbiology to identify new natural products or medicinal chemists to design and make small molecules. The pre-clinical   expertise   needed   to   develop   compounds   with   very   different physiochemical and toxicology profiles is also very broad.   Regulatory requirements are very different from small molecules to biologicals and last, but certainly not least, the manufacturing expertise and capital investment required to produce bulk active ingredient of such a broad range of products is staggering.   Investment into dedicated plants costing up to $250 million can be required to be able to introduce a unique product into the market place.

Given the above factors, it should come as no surprise that one potential scenario for reducing R&D costs should be to find one ubiquitous class of drugs that could address all future unmet medical needs.   This "holy grail" of drug discovery is not really so far-fetched when one considers the antisense oligonucleotide field (1).

The prime assumption of the antisense hypothesis is that all metabolic processes are controlled by proteins which are synthesized by transcription and translation from a DNA template. Antisense compounds interfere with the process of translation of mRNA into protein. Their structures are directly related to the structure of the target mRNA by Watson-Crick base pairing rules.   In principle then, the entire drug discovery process becomes trivial and virtually zero cost i.e. once a target gene or protein is selected, the inhibitor i.e. antisense oligonucleotide analog, can be immediately designed based on the mRNA sequence.   Synthesis of the desired compound can be rapidly accomplished using standard automated nucleotide synthesis and the cycle time for drug discovery from concept to lead compound can be reduced from 5 years and a cost of perhaps to $20 million, to 24 hrs and perhaps a few thousand dollars!

Pre-clinical development can also proceed at very low cost and effort because the physiochemical and non-mechanism related toxicity profile of the drugs will be almost identical! Even though there are theoretically over 1 billion sequence combinations possible for a 20-mer, all of the oligonucleotides will have the same charge distribution, stability and very similar molecular weight! Thus, formulation, metabolism and pharmacokinetics will be routine. Manufacturing of all drugs will be done in a single facility. All that will change will be the sequence of the four bases! Assuming that non-mechanism related toxicities of oligonucleotides are predictable or non-existent, regulatory concerns will be minimal. Indeed, as the predictability of the behavior of these drugs in man grows, regulatory agencies may require less and less animal based toxicity in favor of direct human studies. Paradise indeed! Thus pre-clinical, clinical and manufacturing costs will drop dramatically and drugs could be economically developed for any sub-set of the population, no matter how small.

Fortunately, for thousands of talented chemists and biologists in the pharmaceutical industry, it is not that easy in practice! Firstly, the selection of the optimal antisense oligonucleotide can not simply be inferred from the mRNA sequence but currently has to be experimentally determined by preparing all possible 20 mers, scanning the mRNA sequence (usually some several hundred 20-mers must be prepared and tested) in a process called mRNA oligo-scanning. Secondly, the optimal chemical nature of the antisense oligonucleotides has not yet been determined. The current generation of oligonucleotides are not very stable in biological fluids and there are problems with bioavailability and toxicology. Finally, as the synthetic sequences are comprised of upwards of 100 steps, manufacturing costs are currently extremely high, relative to other drugs.

However, despite the relative infancy of the field, considerable progress is being made in addressing these challenges. Chemistry is making a major contribution to this field with many combinations of phosphodiester isosteres and modified bases being explored to increase the stability and binding specificity of the oligonucleotides (2). In addition, the chemical optimization of the manufacturing processes for synthesizing the oligonucleotides is also making an impact on the cost of producing these drugs.

Standardize the R and D Process

In the previous scenario, the emphasis was on the nature of the product. In the following scenario the emphasis is on the nature of the drug discovery target and the process of finding and developing the resulting product. Currently, each new drug discovery target is pursued essentially as a unique and separate project, with solutions to each problem being developed independently from other projects, without building on previous experience. There are several reasons for this situation. Firstly, until recently, the drug discovery targets available were limited in number and as a result were usually from very different fundamental types of mechanisms e.g. histamine receptors, oxidative enzymes, proteases, calcium channels etc. Thus the probability that any one company would identify a new target in a class where they had previous experience was fairly small. In addition, even when this did occur, the experience was often in a different part of the company (e.g. different therapeutic areas or research sites) and unfortunately, the expertise gained from the previous project was rarely effectively transferred to the new project. Thirdly, the pace of technological advance often meant that previous experience, even should it exist, was out-of-date.

This situation is likely to change dramatically over the next few years as the number of potential new targets increases dramatically due to the worldwide effort to sequence the human genome as well as the genomes of plants, animals and microorganisms (3). As there are likely to be only a limited number of fundamental mechanisms, the likelihood that a newly discovered target will fall into an area of known expertise will increase. Decreased project cycle times and the increasing size of large pharmaceutical companies will also increase the probability of working on a new project that comes from a known family of targets.

The above factors are now making it feasible for companies to focus on the optimization of the processes that are used to identify targets and carry out the drug discovery process. In each of the following segments of the drug discovery process, chemistry is playing a critical role in this new paradigm.

Target Identification.   The process of identifying new drug discovery targets is already being highly optimized by the "brute force" massive sequencing of human genomic DNA and cDNAs derived from mRNA present in cells, and more elegantly, by the use of positional cloning of genes involved in human disease. These techniques rely heavily on advances in sequencing technologies that have been developed using specially designed fluorescent dyes (4) that enable highly efficient and accurate determination of sequences without the use of radioactivity.

In addition, attempts are being made to characterize the human proteome i.e. the set of all proteins expressed in human cells. This technology utilizes 2-D capillary electrophoresis to separate out proteins which are then digested and the resulting peptide fragments identified using mass spectroscopy (5). These physical chemical and analytical techniques when applied together are revolutionizing the study of cellular and molecular biology.

In this segment of the R and D process, the new emphasis is on developing and optimizing processes that very efficiently generate large numbers of potential new drug discovery targets in a short period of time. This approach is in direct contrast to the past where metabolites or proteins, were painstakingly isolated from tissue samples and laboriously characterized, often over a span of many years or even decades.

While the above approach identifies genes and proteins of interest, they do not automatically identify the functions of these proteins. This process of "function discovery" (6) can be trivial if the gene or protein is highly homologous to a known gene, or it can be very time-consuming if the gene has no known homologies and represents a completely new class of proteins. This process of "function discovery" can also be standardized. There are many standard techniques to shed insight into the function of a new gene or its product e.g. electronic sequence searching of databases, protein expression and characterization, determining tissue and cellular distribution, gene knockout or over-expression, yeast two-hybrid system etc. These techniques need to be very efficiently and aggressively brought to bear on the problem of function discovery.

Currently, these techniques are often applied sequentially, often by the discoverer of the gene who learns these techniques as she or he goes along, jealously guarding their gene. This is clearly very inefficient. It is much more efficient to have dedicated groups with the required expertise performing these standard techniques. These can be either in-house support groups or external contract groups, depending on what stage of development the technology is at and how cost effective it is to do the work inside or outside the company.

Once the function of the gene has been determined the traditional drug discovery process is initiated. For the purposes of the following discussion, to exemplify the concept, the example of the discovery of zinc metalloproteinase inhibitors will be used, although the principles are applicable to other examples.

Primary Assay Development. The first step in the drug discovery process is the development of a primary assay. Here the concept of standardization can be readily used to efficiently establish the primary assay. In the case of zinc metalloproteinase inhibitors, the assay should simply be a variation on a theme i.e. a standardized protease assay will be available and all that needs to be changed is the substrate. These assays can be developed by linking a fluorescent donor moiety to a fluorescent acceptor via a peptidic linker consisting of a defined series of amino-acids (7). When the protease cleaves the peptide, a fluorescent signal will be observed. The optimal peptide substrate for each protease can be rapidly found by using combinatorial chemistry techniques to sequentially modify each amino acid at the cleavage site in order to identify the optimal amino-acid sequence. This substrate can then be substituted into the standard, highly automated assay and screening begun. In the future, adapting a standard assay should be the norm, developing a completely new assay should be the exception.

Lead Finding. Again this process has to be tackled in a standard way in order to reduce costs and increase efficiency. For a target with a well known mechanism e.g. a zinc metalloproteinase in this case, an extensive library of all compounds previously made for this class of enzyme should be available. This library should be pre-formatted in 96-well plates and the entire library rapidly screened, perhaps in a matter of days, to generate leads. If no mechanism specific library is available then the high through-put random screening system should be utilized to screen as many as 200,000 compounds rapidly to obtain leads.

If a structure based route is chosen (often in parallel to screening), this should again build on previous experience e.g. a structural model can be built up based on a homologous protein (8) where the 3-D structure has been previously obtained, or the 3-D structure of a related protein can be used to more rapidly solve the X-Ray structure of the target protein.

Lead Optimization. If there is one single area where chemistry has made an impact on the drug discovery process then it is in this area of lead optimization where the new technology of combinatorial chemistry is revolutionizing the field (9). Until recently, the process of lead optimization consisted of medicinal chemists designing, making and characterizing each analog, one at a time, at a rate of only a few compounds per month. This laborious and inefficient approach is being replaced by a process where the medicinal chemist will spend a few months in adapting a synthetic route for use on an automated synthesizer and will then make 500-1000 analogs in the space of a few weeks. All compounds will be rapidly tested, the best analogs selected and the multi-parallel synthesis process repeated to generate the next library of compounds. The field of combinatorial chemistry and automated synthesis is advancing with astonishing speed with the development of new resins, new methodologies for tagging compounds, and for new conditions that can allow standard chemical reactions to be carried out on solid phase.

While it is relatively clear how to standardize the optimization of in vitro potency, it is not so obvious how to optimize the in vivo potency or physicochemical characteristics of compounds. This is an area of active research where for example the measurement and calculation of coefficients such as log P and pKa are being used to develop mathematical models to predict the bioavailability of compounds (*10*).

However, as it is still very difficult to accurately predict the behavior of compounds in complicated biological systems, for the time being , it will probably be sufficient for the above multi-parallel synthesis approach to generate 4 or 5 highly potent analogs which can then be evaluated by pharmacology and pre-clinical development experts to choose the best compound.

Clinical Interface. There are even opportunities to optimize the initial evaluation of compounds in human clinical studies. One of the major stumbling blocks for the development of new compounds is the initial Phase I evaluation of safety and pharmacokinetics in man. There are many examples where, despite extensive pre-clinical analysis of bioavailability in many different animal species, the data derived from clinical trials is completely different, and the compound is dropped after its first exposure to man. Given this lack of correlation between animal and human data, it is logical that getting a compound into Phase I studies should be done as quickly as possible after determining the compound is safe for human exposure, and with the least commitment of pre-clinical and clinical resources.

Thus, the preferred scenario should be that when the selection of a clinical candidate with the required pharmacological profile is made, 1-5 kg of active material should be made, 4 week toxicology carried out, an IND filed and a single dose pharmacokinetic study performed. If the compound does not achieve the desired profile, the bioavailability, pharmacokinetic and metabolism data gained in humans can be quickly fed back to the drug discovery project team and another analog put forward for development which hopefully addresses the faults of the first compound.

In practice, the bottleneck for this clinical strategy is the availability of the 1-5 kg of active material. As a result, most major pharmaceutical companies have made substantial investments in small pilot plant facilities that are flexible enough to run wide ranges of chemical reactions on an intermediate scale in order to obtain the necessary active ingredient. Increasingly, large scale column chromatography techniques are being used to rapidly purify compounds, and combinatorial chemistry methods utilized to optimize reaction conditions to further accelerate the process.

Conclusion. The drug discovery process in the pharmaceutical industry is going through a revolution. In both the "standardize the product" paradigm and the alternative "standardize the process" paradigm outlined above, this revolution is being driven by the development of radical new technologies. In combination with the field of biology, the discipline of chemistry is playing a pivotal role in the emergence and application of these new technologies to the fascinating and vital pharmaceutical industry.

Literature Cited

1. Crooke, S.J.; Leblue B., Eds.; *Antisense Research and Applications*; CRC Press Inc: Boca Raton, Fl, 1992.
2. Kiely, J.S. *Ann. Rep. Med. Chem.* **1994**, *29*, pp 297-306.
3. Lander, E.S. *Science* **1996**, *274*, pp 536-539.
4. Lee, G.L.; Connell, C.R.; Woo, S.L.; Cheng, R.D.; McArdle, B.F.; Fuller, C.W.; Halloran, N.D.; Wilson, R.K. *Nuc. Acid. Res.* **1992**, *20*, pp 2471-2483.
5. Li, G.; Waltham, M.; Anderson, N.L.; Unsworth, E.; Treston, A.; Weinstein, J.N. *Electrophoresis* **1997**, *18*, pp 391-402.
6. Fields, S. *Nature Genetics* **1997**, *15*, pp 325-326.
7. McGeehan, G.M.; Bickett, D.M.; Wiseman, J.S.; Green, M.; Berman, J. *Methods in Enzymology* **1995**, *248*, pp 35-46.
8. Sanchez, R.; Sali, A. *Curr. Opin. Struct. Biol.* **1997**, *7*, pp 206-214.
9. Borman, S. *Chem. Eng. News.* **1997**, *75*, pp 43-62.
10. Obach, R.S.; Baxter, J.G.; Liston, T.E.; Silber, B.M.; Jones, B.C.; MacIntyre, F.; Rance, D.J.; Wastall, P. *J. Pharmacol. Exp. Ther.* **1997**, *283*, pp 46-58.

Knowledge-Driven Research:
New Tools, New Frontiers

8

Trapping Molecules and Liberating Catalysts

Dudley R. Herschbach

Department of Chemistry and Chemical Biology
Harvard University
12 Oxford Street, Cambridge, MA 02138

Molecular beams and lasers have provided versatile tools to probe and tame molecular interactions. This paper, after briefly reviewing means to obtain beams with specified translational energy and rotational orientation, describes prospects for two new techniques that would make accessible unorthodox chemical domains. One seeks to trap molecules in an intense laser field and cool them to microkelvin temperatures. The deBroglie wavelength for relative translation would then become much larger than the molecular size, thereby markedly altering the nature of chemical interactions. The other technique employs supersonic expansion of a reactant gas through a nozzle fashioned from a catalytic material. The gas flow is then far from equilibrium, so large yields can be obtained of products from catalysis that would not survive under equilibrium conditions.

Introduction

In his *Imagined Worlds*, Freeman Dyson emphasizes that new directions in science are launched by new tools much more often than by new concepts.(*1*) He says: "The effect of a concept-driven revolution is to explain old things in new ways. The effect of a tool-driven revolution is to discover new things that have to be explained." This is typical in chemistry. Yet there is also often a reciprocal, symbiotic link between the development of tools and concepts. Here I describe the early stages of two projects in which new tools offer promise of transmuting old concepts into unorthodox and unexplored chemical domains.

One of these projects aims to trap molecules in an intense laser field and cool them to microkelvin temperatures. This would emulate the remarkable development by physicists of ultracold traps for atoms.(2) It would enable study of hyperquantum phenomena in molecular reaction dynamics, as well as the exotic effects already seen for atoms, such as Bose-Einstein condensation. At room temperature, gas molecules typically career about with the speed of rifle bullets, whereas in the microkelvin range they move at the sedate pace of snails.   The deBroglie wavelength thereby becomes comparable to that of electrons. Chemical reactivity will then be governed by quantum tunneling and resonances, so it will differ drastically from that at ordinary thermal conditions.

The other project exploits in a new way a venerable but versatile molecular beam technique, expansion of a gas through a supersonic nozzle. For chemistry this technique, in many variants, has yielded a cornocopica of novel results; foremost among them is the discovery of the fullerenes.(3) The new step is simply to fashion the nozzle from a catalytic material.(4) Since the supersonic expansion is far from an equilibrium process, this markedly changes the consequences of catalysis. According to the traditional definition, a catalyst speeds (or retards) the attainment of chemical equilibrium.   Under supersonic conditions, however, the gas flow is far from equilibrium. The catalytic process then can deliver a host of product species that would not appear in appreciable yield under equilibrium conditions.

As background for the development of these new tools, I will briefly sketch a few pertinent aspects of ancestral molecular beam and spectroscopic methods. Over the past four decades, these methods have enabled chemists to attain increasing control over molecules and photons, in both space and time.(5)   A chief motivation has been to study, under single-collision conditions, the intimate dynamics of molecular interactions.   For the fledgling methods considered here, however, the focus shifts to multicollision conditions and collective behavior, in both molecular trapping and nonequilibrium catalysis.

Taming Molecular Wildness

Ordinarily gas molecules not only dash about in all directions with a broad range of thermal velocities, but also rotate freely with random spatial orientations.   The supersonic beam technique has had a key role in subduing both the translational and rotational wildness of molecular motion.

Collisional Organization of Molecular Beams. Figure 1 contrasts two modes of producing a molecular beam. It pictures a small source chamber within a vacuum apparatus. The canonical physics literature, going back to the 1920s, stressed that the pressure within the source chamber should be kept low enough so that molecules, as they emerge from the exit orifice, do not collide with each other. In this realm of "effusive" or "molecular" flow, the emergent beam provides a true random sample of the gas within the source, undistorted by collisions. Of course, this canonical ideal was blatantly violated when chemists, in the late 1950s, undertook to study reactions in crossed beams. Since such studies desperately needed intensity, much higher source pressures were used. Collisions within the orifice then produce hydrodynamic flow. At high enough pressures, the flow becomes supersonic. That regime intrigued chemical engineers, and by virtue of their work,(6) it was soon recognized that supersonic beams offer marked advantages for collision and spectroscopic experiments.

Supersonic beams have narrow distributions in both direction and molecular speeds. Moreover, the rotational and vibrational temperatures of the molecules can be made very low, of the order of a few degrees Kelvin. By seeding heavy molecules in a large excess of light diluent gas, it is also easy to accelerate the molecules to high

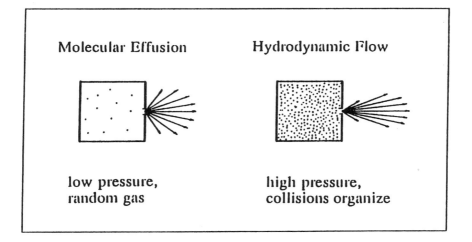

Figure 1. Contrast between production of a molecular beam by effusion (at left) and by a supersonic expansion (at right). Reproduced with permission from reference 29.

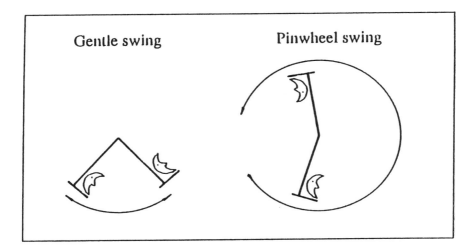

Figure 2. Gymnastic analogy for method or orienting molecules. Reproduced with permission from reference 29.

translational energies, well above typical chemical bond strengths. Suitable seeding can, in addition, induce marked spatial alignment of the rotational angular momenta.

These properties of supersonic beams, all resulting from collisions as the molecules crowd out the exit aperature, are readily understood by Bostonians, but may be less apparent to more polite people. *Filene's*, a famous Boston department store, regularly has Saturday morning sales. Typically, a dense crowd gathers (like the high-pressure gas within the beam source). When the doors are thrown open and the crowd rushes in, collisions induce everybody to flow in the same direction with about the same speed whether they want to or not. Furthermore, if some customers are excited at the prospect of a bargain and leap about or turn handstands (like vibrating or rotating molecules), they suffer more collisions--even black eyes or bloody noses. Thereby such lively customers are calmed down (just as molecular vibration and rotation can be cooled to very low effective temperatures). If a busload of highly kinetic children is turned loose, however, collisions can accelerate the adults to higher velocities (as happens when heavy molecules are seeded in an expanding light gas). The upshot is that, in supersonic expansions, collisions organize the molecular beam far better than could be achieved by any mechanical means.

The customers entering Filene's are not usually spinning or tumbling, but if they were, their collision frequency would vary with the alignment of their angular momenta. A customer cartwheeling parallel to the flow would less often smack into others than would a customer cartwheeling perpendicular to the flow. Recent work has demonstrated the molecular analog of this effect, enhanced by seeding, and other more subtle factors that foster collisional alignment of rotational momenta.(7)

Field-Induced Orientation of Permanent Dipoles. Of more chemical interest is the ability to constrain the spatial orientation of a molecular axis, rather than just its rotational momentum. That capability makes feasible experiments probing the orientation dependence of reaction probability. One means of obtaining beams of oriented molecules, developed in the late 1960s, has provided incisive information about reaction stereodynamics.(8) However, the method is limited to a special class of dipolar molecules that undergoes precessional motion about a space-fixed direction, rather than tumbling end-over-end. Such molecules must have a three-fold or higher axis of symmetry, or equivalent electronic momentum. In a hexapolar electric field, certain of the precessing rotational states can be selected and focused into a beam. This method does not induce orientation but rather selects molecules already oriented in space by virtue of their precessional motion.

For more typical molecules which pinwheel rather than precess, such as linear or asymmetric top rotors, it was long presumed that appreciable orientation could not be obtained by means of an electric or magnetic field. Under ordinary conditions, for pinwheeling molecules the peripheral velocity is very high, again comparable to a bullet. At field strengths that can be maintained in a beam apparatus, even for strongly dipolar molecules the tumbling will merely speed up a bit as the molecule rotates into a favorable orientation for interaction with the field, and slow down a little for unfavorable orientations. Since the molecules continue to pinwheel, however, the net orientation attained is negligibly slight. For over 25 years this situation was lamented in numerous research papers and reviews.(9)

A simple solution, applicable to any dipolar molecule that interacts substantially with an external field, was demonstrated in 1990 and is now used widely.(*10*) Subjecting the molecules to a supersonic expansion drops the rotational temperature quite low before they enter the field. The tumbling is then languid, so in a strong field a molecule can no longer pinwheel. Rather, it swings to and fro about the field direction like a tiny pendulum, and thus is oriented in space. As indicated in Figure 2, this is analogous to a child on a swing going gently back and forth in the Earth's gravitational field rather than recklessly gyrating. It is curious and embarrassing that such an easy method was so long overlooked, even during a period when rotational cooling in supersonic expansions was well understood and exploited to simplify spectra. The method works for precessing as well as pinwheeling rotors. Also, the field required can be supplied by a pair of thumbnail size electrodes; there is no need for hexapolar focusing fields a meter or more long.

Laser-Induced Alignment of Polarizable Molecules. Many molecules lack a permanent electric or magnetic dipole moment and thus cannot be oriented in the way just described. However, since for all molecules the electron distribution is polarizable, an external field can induce a dipole moment to interact with in second-order. For static fields of feasible strength, the induced dipole is too small to matter. But an intense nonresonant laser field can do much better.(*11*) Although the electric field of the laser rapidly switches direction, since the interaction with the induced dipole is governed by the square of the field strength, the direction of the force experienced by the molecule remains the same. The polarizability of a molecule is in general anisotropic (e.g., typically twice as large along as transverse to the axis of a linear molecule). This anisotropy of the induced dipole interaction provides a means to produce pendular alignment of the molecular axis regardless of whether the molecule is polar or paramagnetic or neither. The rapid oscillation of the laser field averages out the interaction with any permanent dipoles, since these depend on the first power of the field strength rather than its square.

The distinction between orientation and alignment should be noted. An oriented vector is like a single-headed arrow (→), an aligned vector like a double-headed arrow (↔). A molecule is oriented when a particular end of its axis points on average in a specific direction, usually that of an external field. This requires that the two ends of the molecule differ, as with a polar molecule. The molecule is aligned when both ends point on average with equal probability in a given direction and its opposite. Whether or not the ends of the molecule differ, only alignment can be attained when the external field reverses direction rapidly. Since the polarizability anisotropy does not distinguish between the two ends of the molecule, the induced dipole is the same in both directions.

Because the polarizability interaction creates a double-well potential, with end-for-end symmetry, the energy levels associated with the librational motion of an aligned molecule are split by tunneling between the potential wells.(*12*) The tunneling rate depends exponentially on a dimensionless parameter proportional to the polarizability anisotropy, the laser intensity, and the reciprocal of the rotational constant of the molecule. Accordingly, the tunneling rate can be varied over many orders of magnitude by scanning the laser intensity over a modest range. This has not yet been done, but in a molecular beam experiment it should be feasible to observe directly the rate of interconversion of "orientational isomers" by such angular tunneling; the process would be akin to the interconversion of enantiomeric molecules.

Pendular alignment by means of the induced dipole interaction is only beginning to be explored.   However, as illustrated in Figure 3, the alignment can be quite pronounced. Striking experimental evidence for such alignment has been found in the angular distributions of fragment atomic ions from multiphoton dissociation of CO and I_2 molecules produced by very intense (10^{14} W/cm^2) nonresonant pulsed infrared lasers.(13,14) Clear evidence for substantial alignment in a much less strong laser field has also been seen in the nonlinear Raman spectra of benzene-argon complexes.(15)

Trapping and Cooling Molecules

The interaction of the induced dipole moment, proportional to the polarizability, with the laser field is purely attractive. Thus in principle the polarizability interaction offers a means to attain spatial trapping of molecules in a fashion similar to a far-off-resonance technique used for atoms.(16) This is an inviting possibility since the "optical molasses" methods typically employed for atom traps do not work for molecules because the oscillator strength of the electronic transition is spread over many vibrational and rotational lines.

To trap an atom or molecule requires not only an attractive potential well, but a means to remove enough kinetic energy so the particle cannot escape from the well. As a whimsical analogy, consider a moth drawn to a candle flame (just as a polarizable molecule will be drawn toward an intense spot of laser light). The moth has enough kinetic energy to fly on by the flame, but often gets so close its wings are singed (a sure way to deprive it of kinetic energy!) and it never escapes. What we want to do with molecules is nicely expressed in an old Quaker benediction: "I'm holding you in the light."

A theoretical assessment of the prospects for this method necessarily involves some guesswork, but the results are encouraging.(4) Estimates of laser fields indicate that by focusing a strong cw infrared laser to a beam waist of about 10 mm diameter, using high-finesse cavities, it may be possible to obtain a laser field of the order of 10^{11} W/cm^2 within a trapping volume of about 10^{-9} cm^3. This is sufficient to enable the induced dipole interaction to provide a well depth of the order of a few °K or more for typical molecular polarizabilities. For instance, for such a field the computed well depths for CO, I_2 and C_{60} molecules are 3, 20, and 120 °K, respectively. To "catch" a substantial population of molecules within the trap, we need to drop the kinetic energy (for rotation as well as translation) to a range substantially below the well depth. Note that the anisotropy of the polarizability does not matter for trapping, but just the net average attractive interaction. If the anisotropy is large, the trapped molecules will be both aligned and confined.

A nice technique for soaking up kinetic energy has worked very well for loading atom traps(17) and appears well suited for the molecular case too. This would dissipate the molecular translational and rotational energy by collisions via collisions with a cold buffer gas. The best choice would be ^{3}He, maintained at about 250 mK by a dilution refrigerator or other cryogenic device. At or above 250 mK the density of ^{3}He exceeds 5×10^{15} cm^{-3}, adequate for collisional relaxation. Since the polarizability of helium is

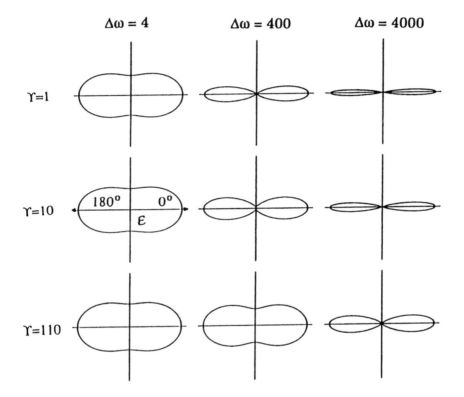

Figure 3. Polar plots of ensemble-averaged alignment angular distributions; $\Delta\omega$ and Ψ are dimensionless parameters involving the anistrophy of the molecular polarizability, rotational constant, laser field strength, and gas temperature. Reproduced with permission from reference 11.

quite small, it will not be trapped. Thus, once the molecules of interest are cooled down by the buffer gas and thereby confined, the buffer gas can be pumped away.

The anticipated mechanism by which the molecules are "held in a pocket of light" is illustrated in Figure 4, to bring out some instructive details. Both the form of the light "pocket" and the spatial distribution of trapped molecules are indicated. The focused laser beam has a hyperboloid shape, characterized by w_0, the minimum waist radius, and by z_R, the axial distance in which the waist expands to $2^{-1/2}w_0$. The effective volume of the trap is defined by the distribution of laser intensity, Gaussian in the radial direction and decreasing linearly in the axial directions away from the trap center. At any point, the trap depth produced by the polarizability interaction is proportional to the laser intensity. Thus, at axial positions $\pm z_R$ from its center the trap is shallower by a factor of two as well as wider by a factor of $2^{1/2}$. The spatial distribution of trapped molecules, governed by the Boltzmann factor for the ratio of local trap depth to temperature, accordingly is Gaussian in the radial direction but its height and width vary with the axial location. For the example shown, at $\pm z_R$ the peak population is 50-fold smaller than at the trap center, the bottom of the "pocket." Although the light beam itself is "open at both ends," the Boltzmann factor amplifies the weakening of the polarizability interaction that occurs away from the beam waist and thereby in effect closes off both "ends of the pocket."

Once the molecules are confined in the trap, at the temperature of the buffer gas, the game has just begun. Provided the number densities are sufficiently high to allow collisional equilibration within the trap, the temperature of the trapped ensemble can be lowered much further by means of evaporative cooling. This technique, well established for atom traps, (18) is an iterative process. It is applied after the buffer gas has been pumped out. Lowering the trap depth by decreasing the laser intensity permits the molecules with the highest kinetic energy to escape. Subsequent rethermalization of the remaining imprisoned molecules then drops their temperature. Tightening the trap by turning up the laser intensity can raise their density markedly. If the process works for molecules as well as it has for atoms, final temperatures in the microkelvin range can be expected.

I must emphasize that the pursuit of molecular trapping is a sporting proposition. As usual in discussion of prospective experiments, I've not dwelled on things that might prove major roadblocks. For instance, under trapping conditions, molecules may be much more likely than atoms to aggregate and simply condense out as fog or snow. Indeed, the success of atomic traps seems uncanny, particularly in attaining temperatures low enough and densities high enough to allow observation of the Bose-Einstein transition in the gas-phase. For 70 years, Bose-Einstein condensation was presumed to be attainable only for liquid Helium. That was because for the conditions required to form a degenerate quantum gas, thermodynamics would impose the mundane process of ordinary condensation.(19) Atomic traps evade thermodynamic equilibrium by means of drastic kinetic inhibition of ordinary condensation. This is done by making the temperature extremely low before building up the density, so the rate of approaching equilibrium remains very slow. This extent to which this might work for molecules has to be found out experimentally.

Also, I should mention that "trapping," although now a customary terminology, is ridiculously inappropriate. The tightest traps in prospect have volumes of the order of femtoliters, so their linear dimensions are at least about 10^4 Angstrom units. For

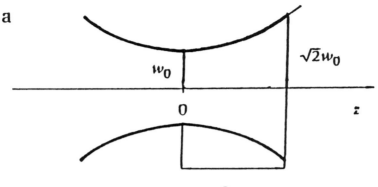

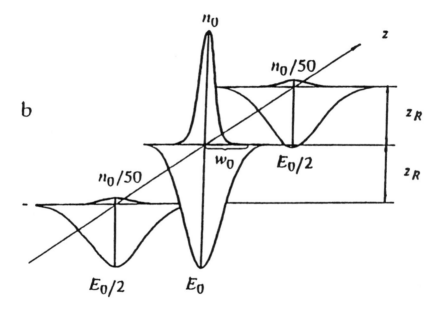

Figure 4. Form of focused laser beam and spatial distribution of trap depth and molecular population. (a) Hyperboloid shape of laser beam; w_0 denotes the minimum waist radius, z_R the axial distance in which the waist radius expands to $2^{1/2}w_0$. (b) Interaction energy (plotted downwards), computed for trap produced by polarizability interaction for CS_2 molecules at 0.24 K, subjected to focused laser of intensity 2×10^{10} W/cm2 and wavelength = w_0 about 10 μm. For this case, the maximum trap depth E_0 is −2 K and the peak density n_0 of trapped molecules if 5×10^{10} cm^{-3}. Reproduced with permission from reference 30.

typical atoms or molecules that's by no means a cage, but just a rather roomy corral. However, even if the molecules range about randomly within such a corral, that degree of confinement can be exploited in many ways, particularly since they will roam leisurely, like snails rather than bullets or even horses. There are also possibilities for organizing molecular movements within the corral. To avoid roaming too far into fantasy, here I merely note a kindred scheme (not necessarily involving trapping), recently proposed.(20) This offers a practical design for a molecular storage ring, although the orbiting molecules would be galloping rather than creeping along.

Prospects for Hyperquantum Chemistry. Beyond the striking phenomena that atomic traps have made accessible for study, molecular traps offer access to a strange new regime for study of chemical reactivity. At first blush, interacting gaseous molecules at microkelvin temperatures would seem to preclude any appreciable reaction. Even a minuscule activation energy would greatly exceed the available thermal collision energy. However, when the translational velocity becomes very low, the deBroglie wavelength can become large compared with the size of the molecules. This entirely changes the nature of reaction dynamics.

For molecules of mass M at thermal equilibrium at temperature T, the deBroglie wavelength corresponding to the average thermal momentum is given by $\Lambda = h/(2\pi M k_B T)^{1/2}$, where h is Planck's constant and k_B the Boltzmann constant. For Λ in Angstrom units, M in g/mol, and T in degrees Kelvin, $\Lambda = 17.4/(MT)^{1/2}$; this is displayed in Figure 5 as a nomogram. Thus, at room temperature a hydrogen atom has a thermal deBroglie wavelength of about one Angstrom unit (10 nanometers in current textbook parlance). But at 10 millikelvin, a chlorine molecule has $\Lambda = 20$ Å; at 1 microkelvin, $\Lambda = 2000$ Å. At such low temperatures, collisions of even fairly heavy molecules exhibit very marked quantum effects.

When the wave character of the molecular translational motion is so pronounced, a reactive encounter becomes something like the coupling of amoebas. Rather than picturing the transfer of a ball-like atom between localized regions of the reactant species, we need to visualize reactant waves mixing, sloshing or oozing about, then gurgitating forth the product waves. The familiar concept of a potential energy surface still pertains, but it is no longer appropriate to think of particle trajectories traversing it; rather what matters is how the surface slows or speeds the progress of surging waves.

If as usual the potential surface has a substantial activation barrier, in the hyperquantum regime tunneling becomes the dominant reaction pathway. When Λ is much larger than the thickness of the barrier, the rate constant is governed by a threshold law derived by Wigner.(21) Accordingly, for an exothermic bimolecular reaction, below some low temperature, reaction occurs entirely by tunneling, and the rate does not change with a further drop in temperature. If the barrier thickness for the reactants in their ground internal states is denoted by D and the reduced mass for the collision partners by μ, the criterion $\Lambda \gg D$ requires $T \ll 300/\mu D^2$. Hence, for a generous estimate $D \approx 10$ Å, to attain the Wigner limit the temperature needs to be well below one degree Kelvin for a hydrogen atom reaction and about a hundredfold lower for heavier reactants like chlorine molecules.

Of course, at ordinary temperatures, tunnel effects are often appreciable for hydrogen atom reactions. That is even more so for reactions of muonium; indeed, for

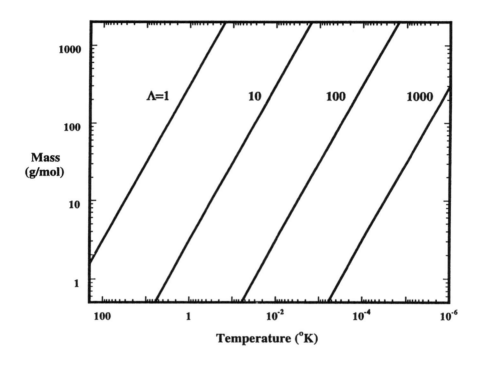

Figure 5. Nomogram for thermal deBroglie wavelength Λ (in Å units) as a function of molecular mass (g/mol) and temperature (°K).

the Mu + F$_2$ reaction nice evidence has been found for a transition towards the Wigner threshold tunneling in the gas phase.(22)   The low-temperature limiting behavior expected when tunneling is dominant has also been seen for several reaction processes in the solid phase.(23) Accordingly, at least for exothermic bimolecular reactions with activation barriers, such results indicate the general nature of the tunneling behavior that can be anticipated in an ultracold molecular trap. For the study of tunneling processes, the trap environment offers a great advantage, however, since the imprisoned molecules can keep colliding with each other until all have undergone the tunneling reaction.

Whether or not the potential energy surface has an activation barrier, when Λ is large compared with the size of the molecules the orbital angular momentum in reactive collisions will usually have to be zero (s-wave scattering). In the usual semiclassical picture, this would imply a "head-on" encounter, but that does not pertain when waves collide. Any nonzero angular momentum will inhibit reaction by imposing a centrifugal barrier which will deflect the reactant waves or act to require or to slow a tunneling process.

If the potential surface is purely attractive, with no activation barrier, the cross section for collisions with zero angular momentum becomes, in the long wavelength limit, simply 4π times the square of a quantity called the scattering length. Since this corresponds to the surface area of an effective scattering sphere, it exemplifies how, like an amoeba, the wave engulfs the target. The effective radius of the sphere, given by scattering length, is governed in the long wavelength ("zero-energy") limit by resonances with low-lying virtual or bound states of the collision complex.(24)

This quick prospectus has noted only the most obvious aspects of hyperquantum chemical dynamics. But tunneling and resonances are characteristic of this regime, in that such quantum effects serve as very sensitive probes of particular features of the potential energy surface.(25) For many other phenomena, not involving reactions but interference of coherent matter waves, hyperquantum molecular vapors can offer options not available with atoms. Although it is still quite uncertain if and when molecular trapping will be achieved, one prediction seems safe. In view of the history of the laser and other versatile scientific tools, the chief applications of molecular trapping most likely will not be among those so far contemplated.

Liberating Catalysis from Equilibrium Constraints

Way back in 1912, in his classic work on the interaction of hydrogen with tungsten, Langmuir described a prototypical heterogeneous-homogeneous reaction, in which radicals formed by dissociative chemisorption desorb into the gas phase and react there. In modern times, the role of such surface-generated gas-phase radicals in catalytic reactions has been directly demonstrated in several processes and suspected or postulated in many more. Since typically surface chemistry is studied under equilbrium or near-equilibrium conditions, however, the observable products usually are determined by thermodynamics rather than kinetics.

Recent exploratory experiments demonstrate that a dramatically different regime becomes readily accessible by simply expanding a reactive gas through a nozzle made from a catalytic metal.(4) Thus, on flowing ethane gas at about 80 torr through a nickel nozzle heated to about 1000 °C, about 40% is converted to higher hydrocarbons, C_nH_m. These hydrocarbons contain from n = 3 to 12 carbons and a range of hydrogen

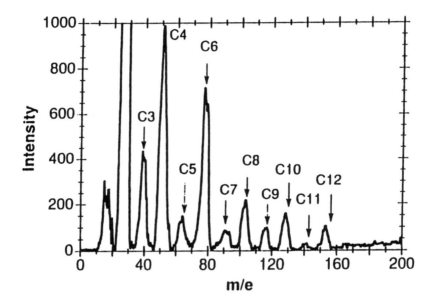

Figure 6. Low resolution mass spectrum (from Ref. 4) of products from Nicatalyzed reaction of ethane (1060 °C, 85 Torr) expanded from supersonic nozzle. Product peaks for C_nH_m are labeled by n, the number of carbon atoms. Arrows indicate masses for m = n (located at n x 13 mass units; thus m = n up to n = 8 but the most probable m is lower by 1 to 3 units for higher n. Reproduced with permission from Elsevier Science and American Chemical Society (reference 4).

atoms, mostly equal to or near the number of carbon atoms (m ≈ n). The n = 6 product appears to be largely or solely benzene, formed in about 15% yield. Figure 6 shows a low-resolution mass spectrum, to give an overall impression of the product distribution.

In contrast, at thermodynamic equilibrium ethane at 1000 °C would decompose entirely, to form carbon on the metal surface and hydrogen in the gas phase. The remarkably facile formation of higher hydrocarbons in a supersonic flow thus exemplifies how kinetics can be exploited to evade thermodynamic limitations. The contact time for gas within the nozzle is about 10ms. Moreover, the strong flow likely fosters desorption of surface species into the gas-phase, thereby further restricting the interaction of reactive intermediates or products with the surface. Much or most of the association reactions producing the higher hydrocarbons probably occur during the free jet expansion just beyond the nozzle exit.

By virtue of the simplicity of the experimental method, chemical variations are easily examined. Replacing ethane by C3 or C4 alkanes or alkenes produces a substantial increase in the yield of higher hydrocarbons, with up to about 45% benzene, whereas with methane no appreciable net reaction is observed. If the nozzle is made from iron or stainless steel rather than nickel, virtually no higher hydrocarbons are formed. Many other combinations of catalytic surfaces and gaseous reactants invite study, and pressure, temperature, and contact time within the nozzle can be varied widely. The recent development of techniques(26) capable of monitoring surface species during reactions at high pressures and temperatures should encourage more incisive investigations of surface reactions with flowing gas.

Benedictory Comments

Science lives at its frontiers. New tools, experimental or conceptual, thereby have a vital role. The two exploratory projects I've described, trapping molecules and liberating catalysts, offer a cultural contrast; one stems from sophisticated physics, the other from naive but robust chemistry. The contrast is meant to give some impression of the grand scope of modern physical chemistry. This great scope becomes still more apparent in the host of other outposts on the current frontier that employ lasers and supersonic molecular beams in quite different ways. Particularly striking are the many variants of femtosecond spectroscopy(24) and other means of probing transition states(27) and work enabling the laser control of reaction pathways.(28)   For physical chemistry this is an exhilerating era, by virtue of the abundance of powerful new tools now in hand and in prospect.

Acknowledgments

I am grateful for the opportunity to work with extremely able and evangelical students and colleagues in pursuit of the projects described here, especially John Doyle, Bretislav Friedrich, Manish Gupta, Daniel Katz, and Lina Shebaro. The work on taming and trapping molecules has been supported chiefly by the National Science Foundation; that on liberating catalysts by Molten Metal Technology.

Literature Cited

1. Dyson, F. *Imagined Worlds;* Harvard University Press: Cambridge, Mass., 1997, pp 49-53.

2. Levi, B.G. *Physics Today,* Dec., **1997**, *50*, p 17 gives a nice account of the saga of atom cooling, tracing the work of C. Chu, C. Cohen-Tannoudji, and W.D. Phillips.

3. Baum, R. *Chem & Eng. News,* **1997**, *75*, (29). Jan. 6 and Feb 17, 1997 give an overview of the fullerene saga, featuring the work of R. Curl, H. Kroto, and R. Smalley.

4. Shebaro, L.; Abbott, B.; Hong, T.; Slenczka, A.; Friedrich, B.; Herschbach, D. *Chem. Phys. Lett.* **1997**, *271*, p 73; Shebaro, L.; Bhalotra, S.R.; Herschbach, D. *J. Phys. Chem.* **1997**, *101*, p 6775.

5. For collections of representative papers, see: *J. Phys. Chem.* **1991**, *95*, pp 7961ff ; **1993**, *97*, pp 2038ff and 12423ff; **1997**, *A 101*, pp 6339ff.

6. For a charming review, see Fenn, J.B. *Ann. Rev. Phys. Chem.* **1996**, *47*, p 1.

7. Pullman, D.; Friedrich, B.; Herschbach, D. *J. Phys. Chem.* **1995**, *99*, p 7407 and work cited therein.

8. For reviews, see Brooks, P.R. *Science* **1976**, *193*, p 11; Parker, D.H.; Bernstein, R.B. *Ann. Rev. Phys. Chem.* **1986**, *40*, p 566; Bulthuis, J.; van Leuken, J.J.; Stolte, S. *J. Chem. Soc. Faraday Trans.* **1991**, *91*, p 205.

9. See, for example: Bernstein, R.B.; Levine, R.D.; Herschbach, D.R. *J. Phys. Chem.* **1987**, *91*, p 5365.

10. For reviews, see: Loesch, H.J. *Ann. Rev. Phys. Chem.* **1995**, *46*, p 555; Friedrich, B.; Herschbach, D. *Int. Revs. Phys. Chem.* **1996**, *15*, p 325.

11. Friedrich, B.; Herschbach, D. *Phys. Rev. Lett.* **1995**, *74*, p 4623; *J. Phys. Chem.* **1995**, *99*, p 15686.

12. Friedrich B.; Herschbach, D. *Z. Phys. D,* **1996**, *36*, p 221.

13. Normand, D.; Lompre, L.A.; Cornaggia, C. *J. Phys. B.* **1992**, *25*, L497.

14. Dietrich, P.; Strickland, D.T.; Laberge, M.; Corkum, P.B. *Phys. Rev. A* **1993**, *47*, p 2305.

15. Kim, W.; Felker, P.M. *J. Chem. Phys.* **1997**, *107*, p 2193.

16. Miller, J.D.; Cline, R.A.; Heinzen, D.J. *Phys. Rev. A* **1993**, *47*, R4567.

17. Doyle, J.M.; Friedrich, B.; Kim, J.; Patterson, D. *Phys. Rev. A* **1995**, *52*, R2525.

18. Doyle, J.M.; Sandberg, J.C.; Yu, I.A.; Cesar, C.L.; Kleppner, D.; Greytak, T.J. *Physica B* **1991**, *194*, p 13 and references cited therein.

19. See, for example, the discussion given by Schrödinger, E. *Statistical Thermodynamics;* Cambridge University Press: 2nd. Ed., 1952, p 54.

20. Katz, D.P.; *J. Chem. Phys.* **1997**, *107*, p 8491.

21. Takayanagi, T.; Masaki, N.; Nakamura, K.; Okamoto, M.; Sata, S.; Schatz, G.C.; *J. Chem. Phys.* **1987**, *86*, p 6133; **1989**, *90*, p 1641.

22. Baer, S.; Fleming, D.; Arseneau, D.; Senba, M.; Gonzalez, A. *Isotope Effects in Gas-Phase Chemistry,* J.A. Kaye, Ed.; ACS Symposium Series No. 502; American Chemical Society: Washington, D.C. 1992, Chap. 8, pp 111-137. See also Fleming, D.G., et al *J. Chem. Phys.* **1989**, *91*, p 6164.

23. Benderskii, V.A.; Makarov, D.E.; Wight, C.A. *Chemical Dynamics at Low Temperatures;* John Wiley & Sons: New York, NY, 1994; Sims, I.R.;   Smith, I.W.M. *Ann. Rev. Phys. Chem.* **1995**, *46*, p 109.

24. Zewail, A.H. *J. Phys. Chem.* **1996**, *100*, p 12701.

25. See, for example, Kuppermann, A.; Wu, Y.M. *Chem. Phys. Lett.* **1995**, *241*, p 229.

26. Somorjai, G.A. *Z. Phys. Chem.* **1990**, *94*, p 1432.

27. Simpson, W.R.; Pakitzis, R.P.; Kendel, S.A.; Lev-On, T.; Zare, R.N. *J. Phys. Chem.* **1996**, *100*, p 7938.

28. Gordon, R.J.; Rice, S.A. *Ann. Rev. Phys. Chem.* **1997**, *48*, p 601; Zare, R.N. *Science* (in press) and work cited therein.

29. Herschbach, D.R. "The Shape of Molecular Collisions," in *Science and Society*; Moskovits, M., Ed; House of Anansi Press, Ltd: Concord, Ontario, 1995; (John Polanyi Festival Toronto, Canada), pp 13-28.

30. Friedrich, B.; Herschbach, D.R. *J. Chinese Chem. Soc.* **1995**, *42* pp 111-117. ("Spatial Taming and Trapping of Molecules")

The Chemical Industry:
Research for Competitiveness

9

Vision 2020—The Chemical Industry

Paul S. Anderson

The DuPont Merck Pharmaceutical Company
P.O. Box 80500, Experimental Station, Wilmington, DE 19880-0500

Productivity and competition are critical issues for chemical industry in the United States. To be competitive on a global scale, industry must produce goods and services that have real value for its customers at a competitive price under environmentally sound conditions. Productivity of this type requires continuous innovation on an increasingly short time scale. It is important for academe, government, and industry to understand this fact and to develop shared values that will insure a high level of productivity for the chemical enterprise.

The Productivity of the Chemical Industry in the 20th Century

Fifty years ago, at the end of the second World War, the United States began to build a science and technology platform that could insure the economic growth and prosperity of this country. The concept as set forth by Vannevar Bush and others was simple. A significant federal investment in science and technology would yield large dividends in terms of new businesses and employment opportunities. The dream, of course, was realized. Through a partnership between the federal government, academe, and industry — with the federal government playing a necessary and critical role — we built a powerful research enterprise. Federal support for research in the university system and the national laboratories produced new knowledge and a cadre of highly-skilled, technically-trained scientists and engineers. Industry provided the opportunities for this skilled workforce to use new knowledge to create new products. The end result was an economic engine that drove our economy and propelled the United States into global leadership.

The U.S. chemical industry is an example of the federal government, academe, and industry partnership. This industry directly employs over one million people including nearly 100,000 scientists and engineers. It is a keystone industry because its products enable other industries to function and employ additional millions of workers. Chemical industry produces approximately $400 billion of finished products each year. Overall, this industry is the third largest manufacturing sector in the nation, representing approximately 10% of all U.S. manufacturing. Over the past decade it has produced a positive balance of trade between $15 and $20 billion per year. While chemical industry invests nearly $18 billion per year in research and development, it has benefited from the federal investment in research and development. The total federal investment in chemistry-related research is approximately $300 million per year. That modest investment is multiplied many fold by the chemical industry.

More than 9,000 corporations develop, manufacture, and market chemical products and processes. Chemicals are a keystone of U.S. manufacturing, essential to most other industries such as pharmaceuticals, automobiles, textiles, furniture, paint, paper, electronics, agriculture, construction, appliances, and services. Chemicals, therefore, are building blocks for products that meet our most fundamental needs for food, shelter, clothing, medicines as well as the products vital to our worldwide communications and transportation systems.

Technology Vision 2020: The U.S. Chemical Industry

At this point it is appropriate to ask what the chemical enterprise should do to insure that it will be at least as productive in the 21st century as it has been in the 20th century. We need to identify the challenges and opportunities that are likely to shape our future. Many of the relevant issues are addressed in a report called *Technology Vision 2020: The U.S. Chemical Industry.*(1) The development of this report began with a request from the White House Office of Science and Technology Policy. This request sought input on how the U.S. government could better allocate R&D funding to advance the manufacturing base of the U.S. economy and insure competitiveness in global markets. The American Chemical Society decided to take the lead in responding to this request. The Society asked the American Institute of Chemical Engineers, the Chemical Manufacturers Association, the Synthetic Organic Chemical Manufacturers Association, and the Council for Chemical Research to join in a joint effort to create what we now call *Vision 2020: The U.S. Chemical Industry.* It is the first report of its kind produced by the chemical industry. Over 200 technical and business leaders in the U.S. chemical industry worked together to study the factors that are affecting the competitiveness of our industry in a rapidly changing business environment. The report focuses on these factors and the technology that is needed to maintain industry's competitive advantage while protecting and improving the environment. The key factors affecting competitiveness were identified as being:

- Societal demands for environmental sustainability.
- Higher customer expectations for product performance and value.
- Workforce requirements for skilled innovative employees.
- Increasing globalization of markets.
- Financial markets requiring increased profitability and capital productivity.

We will come back to a brief discussion of these factors. Let me first summarize the vision statements from *Vision 2020* that describe goals for the U.S. chemical industry as it seeks to preserve its global competitive advantages well into the next century.

- The U.S. Chemical Industry leads the world in technology development, manufacturing, and profitability.
- The U.S. Chemical Industry is responsible for breakthroughs in R&D that enhance the quality of life worldwide by improving energy use, transportation, food, health, housing, and environmental stewardship.
- The U.S. Chemical Industry leads the world in creating innovative process and product technologies that allow it to meet the evolving needs of its customers.
- The U.S. Chemical Industry sets the world standard for excellence of manufacturing operations that protect worker health, safety, and the environment.
- The U.S. Chemical Industry is welcomed by communities worldwide because the industry is a responsible neighbor who protects environmental quality, improves economic well-being, and promotes a higher quality of life.
- The U.S. Chemical Industry sets the standard in the manufacturing sector for efficient use of energy and raw materials.
- The U.S. Chemical Industry works in seamless partnerships with academe and government creating "virtual" laboratories for originating and developing innovative technologies.
- The U.S. Chemical Industry promotes sustainable development by investing in technology that protects the environment and stimulates industrial growth while balancing economic needs with financial constraints.

Achieving Sustainable Living through Chemistry

These, of course, are vision statements. Commitment, action plans, and R&D effort and expenditure will be required to achieve this vision. In many cases, we will need to reinvent how chemicals and materials are produced and used. We will need to look beyond controlling waste at the end of a process and instead study the entire life-cycle of chemical production so that new ways can be discovered to more efficiently produce useful products with less waste or preferably no waste. As we all know, unsaleable by-products (waste) are common in chemical production. A less visible but equally important problem is the large amount of energy that is consumed in manufacturing chemicals. New technology is needed for more energy efficient production of chemicals and useful products derived from them. New chemistry is needed for efficient conversion of sustainable biomass into creative new products and for more efficient generation of existing products from petroleum feed stocks. Production efficiency also depends on proper management of the supply chain and careful engineering of the production facility. These are activities that require extensive computer modeling for achievement of optimal efficiency. Each of these objectives represents an intellectual challenge and an opportunity for innovation. We must develop the science and technology that will allow us to consume at a rate that is in balance with what earth can produce and dispose of waste in a manner that is in balance with what the earth can process and assimilate. We must become convinced that sustainable living is an achievable goal that makes economic sense as we use chemistry to innovate for a cleaner, safer, healthier world.

It is here that U.S. companies have a real competitive advantage. We lead the world in the rapidly growing business of environmental technologies. These technologies will reduce risk, enhance cost effectiveness, and improve process efficiency by creating products and using processes that are environmentally beneficial or benign. A company that can provide a quality product with the lowest energy and raw material consumption and lowest generation of unsaleable material will have a competitive advantage. This is a basic business concept that is tied to productivity and customer expectations.

It is clearly stated in the Chemical Manufacturers Association Program called Responsible Care. In Responsible Care, member companies of CMA strive to improve performance in health, safety and environmental quality for manufacturing processes and to communicate this effort to the public while listening to and responding to public concerns.

Innovation as a Means of Business Survival

Everywhere today customers look for value. We do that when we buy food, clothing, cars, appliances, etc. We are more focused on value than any society in history because we advertise it all the time. My product is better than yours and this is why. We do this because there are more products than customers. Business knows this and it knows that success depends on getting customers and retaining them. One way to do this is to constantly improve an existing product or make it more cheaply or replace it with a superior product. In the chemical industry this could mean new synthetic methods, better process engineering, new catalysis, or discovery of a superior material or a molecule with new properties. Whatever it is, it surely means innovative research and development. This brings us to the work force. If you are going to depend on innovation as a means of business survival, you better have access to new science and technology and innovative people who can figure out how to use it to achieve your business plan. When you look at successful companies in a rapidly changing business world that is innovation driven, you quickly realize that the best ones have the best people. They understand the needs of their customers and then use the creative talent of their employees to make new products that will better serve these needs. Frequently the advances are incremental in nature, but none the less require new science or technology. We are all familiar with the changes in how we listen to music at home. We progressed from phonograph records to tapes to CD recordings. Each step was a progression in terms of sound quality and of easy storage. Each step required an advance in technology.

The outstanding success of the chemical industry is largely due to scientific and technological breakthroughs and innovations brought about through R&D. In fact, the Institute for the Future notes that the chemical industry is one of the eight most research-intensive industries. It is pretty clear that growth and competitive advantage of our chemical industry are linked to individual and collaborative efforts of industry, government, and academe working together to improve the nation's R&D enterprise. Companies that depend on R&D for their future will invest in countries where they have access to skilled people, new knowledge, and a favorable operating climate. That is one of the reasons that the science-based venture capital business is largely a U.S. phenomenon. In some states we have created a science and business environment that strongly supports investments in early stage companies. These companies are very important to the chemical enterprise because they are pushing the frontier of science and technology in a direction of fast paced innovation. By partnering with established companies they can help fulfill our need for new products.

The Chemical Enterprise — A Global Enterprise

The chemical enterprise is, of course, a global enterprise. Large chemical companies operate in many countries. They can invest their capital wherever they please and they will usually do so wherever the return is best. This will undoubtedly become a major problem for us as major markets for chemical products grow in other parts of the world such as Southeast Asia. Most of our large companies are publicly held. Substantial portions of these companies are owned by the investment community. This community likes the number 25. They like to see the sum of growth and return adding up to 25% per year. That is not an easy number to achieve in all sectors of the chemical enterprise. It is certainly an easier number to achieve in parts of the world where costs are low. It is this combination of competition, globalness, and financial expectations that creates significant challenges for *Vision 2020*. It is innovation and the power to innovate that creates the opportunities by which this vision can and I believe will be achieved.

To remain competitive in a capital-intensive global enterprise, chemistry-based companies must be able to innovate. They must deliver creative products that have good value at a competitive price under environmentally sound conditions. This requires highly motivated, well-trained R&D organizations and new chemical knowledge. Fortunately, since World War II, federal dollars have been invested to build and sustain a world-class academic system, with supporting national laboratories, that can and has fulfilled these requirements. Our challenge in days of curtailed federal spending and reduced budgets is to articulate to Congress and the Administration the value that comes from academe, industry, and government working together to achieve national goals that require chemistry and chemical engineering input.

Federal Support for Research

Academe and industry need to jointly advocate the importance of federal support for academic research and training in chemistry as a vital component of what is needed to stimulate economic growth and maintain U.S. competitiveness in global markets. The message is straightforward. Government looks to industry for job creation and economic stimulus. The chemical industry can deliver jobs and economic growth, provided that investment opportunities are favorable, sound science-based environmental policies are in place, a well-trained workforce is available, and there is a flow of new chemical knowledge. A healthy academic system can deliver the workforce and new knowledge. To do that, it must have adequate federal support for its research and training efforts.

Over time, we have created an effective partnership among government, industry, and the academic world that has allowed the U.S. chemical enterprise to achieve competitive leadership in relevant global markets. We need to sustain and nurture this partnership as we move toward the 21st century. As one step toward this goal, the American Chemical Society is strongly supporting efforts to double the federal investment in science and technology over the next ten years. We are leading a coalition of more than one hundred organizations representing over one million scientists, engineers, and mathematicians in the drive to secure our future economic health and prosperity through our national investment in scientific research.

The percent of the Federal budget spent on nondefense-related research and development has been declining for many years. We spend a smaller percentage of our Gross Domestic Product on nondefense research and development than either Germany or Japan. We need to return to a competitive level of investment in science and technology that can fuel growth of new industry. The importance of Federal research dollars for the creation and evolution of the biotechnology industry is well known. This industry, of course, has a strong dependence on chemistry. There now are over 1500 companies in the United States that were founded to commercialize biotechnology. Their expenditures on innovative research and technology development are approaching $10 billion per year and the value of the resulting medical and agricultural products far exceeds this number and is growing rapidly. This is but one example of why the federal investment in research is so critical to our economy and why we in the research community need to be united in our advocacy for this position.

Literature Cited

1. *Technology Vision 2020: The U.S. Chemical Industry*; The American Chemical Society, American Institute of Chemical Engineers, The Chemical Manufacturers Association, The Council for Chemical Research, and The Synthetic Organic Chemical Manufacturers Association; Washington, D.C.; 1996

10

Challenges and Opportunities for Chemical Research in the 21st Century: An Industrial Perspective

Francis A. Via

Director of Contract Research, Akzo Nobel, Inc.
Dobbs Ferry, NY 10522

The accelerated pace of growth in scientific knowledge and of technology evolution in combination with global economics are creating new challenges for chemical industry research and development. These interrelated factors are influencing the character of research in all sectors of our enterprise: government, academia and industry. A greater emphasis is being placed on metrics, productivity, timing, value, knowledge integration, reduced risk and higher returns. Over the last 20 years, the character and organization of industrial research have evolved to a fully integrated system that is a highly responsive component of the corporation it serves. The basis of this publication is to define these influential factors and to demonstrate that chemical research, an enabling technology, will thrive in the 21st Century and will enhance its value with increased global teaming among the three sectors of the enterprise: government, academia and industry.

Introduction

The United States is enjoying the healthiest economy in its entire history and perhaps the strongest of any nation ever. Several hundred thousand jobs are being created each month. It is with great pride that the chemical industry acknowledges its key role in this robust expansion of our economy. In the eyes of many, the chemical industry serves as a strong technical foundation for this prosperity. The U.S. chemical research infrastructure is serving as an important engine for the sustained growth of our nation's industries. These are, indeed, "the best times and the most challenging of times" for many facets of the economy and, in particular, for chemical industry research. For many chemists and chemical engineers, the 80's and 90's have proven to be rather perplexing as conflicting issues have emerged. Technology is one of the principal drivers of our economy, yet the changes and challenges confronting research and development appear to be ignoring its demonstrated high value. Investments in chemical research and development are subject to the same issues as any other investment and these issues have encouraged re-engineering, downsizing, metrics and improved productivity. What is in store in the near future, and how do we maintain the vitality of industrial chemical research?

These are a few of the issues we will attempt to address beginning with a brief overview of the impact of the chemical industry on the U. S. economy and on industrial research. We will set the groundwork for anticipating future activities and needs by examining key industrial drivers, such as the emergence of a global economy and the subsequent globalization of research. Key financial drivers of our industry, such as the equity markets, business trends, liability and regulation are influencing investments in research. Current trends in decentralization and entrepreneurship provide mechanisms for improved accountability and responsiveness with a metric system that fosters a portfolio of short term research programs. Hopefully, it is becoming quite obvious that we are all in this together and that effective integration of industry, government and academic research is underway and is essential for continued economic growth.

The evolution of our technology driven economy is becoming more personally challenging as history has demonstrated, e.g., diesel locomotives, integrated circuits, the plastics industry, magnetic data storage, radial tires and, most recently, digital imaging. Each of these technical transitions is beneficial for society and the economy but, unfortunately, has caused both corporate and personal disruption in its wake. To survive, today's flexible, responsive corporations must master challenges by using and promoting a vital, national, research enterprise.

Outline

To serve as a guide for this survey, an outline of the topics is listed.

> The Chemical Industry - economic impact
>
> Industrial Drivers for Change in the 80's & 90's
> > Globalization
> > Equity Markets
> > Risk & Regulation
> > Chemistry - Enabling or Mature
>
> Chemical Industry R & D - A Profile
>
> Globalization of R & D
>
> Current R & D Trends
> > Innovation
> > Incremental Development
> > Partnerships
> > Global Partnerships
>
> The Future Team

The Chemical Industry

The contributions and economic impact of the U.S. chemical industry as well as that of the chemical research community is frequently unrecognized outside of our small community.   The U. S. chemical industry is the world's largest manufacturer of chemical products by a substantial margin. With a total production of $367.5 billion in 1995; the U.S. chemical industry is 40% larger than that of Japan, ($255.1 billion), and even larger than Germany, ($125.5 billion), France ($85.4 billion), United Kingdom ($61.6 billion) and Italy ($51.0 billion) combined.(*1*)   Most notably, the chemical industry is a leading U.S. manufacturing industry.(*1*)   At 1.8% of the gross national product (GNP), chemicals edges out food and related products at 1.7%, and clearly leads electronics, 1.5% and motor vehicles, 1.1%, see Fig. 1.   Another significant asset of the chemical industry is its positive contribution to the otherwise troublesome area of U.S. balance of trade.   In 1996, the chemical industry was #1 with $61.8 billion in exported products, just edging out the agricultural industry at $61.2 billion and surpassing the much publicized aircraft industry of $40.5 billion by 50%.(*1*)   The balance of trade for the chemical industry was positive by about $17 billion. This comparison is not meant to foster a competition with these industries which are important customers for chemical products, but rather to focus attention on the leading role of the chemical industry in our economy.

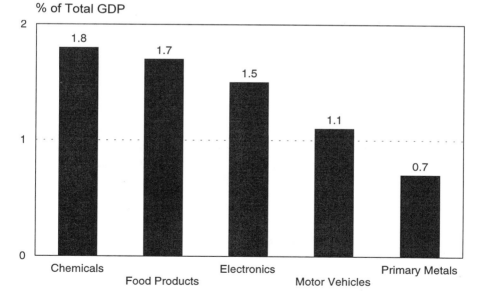

Figure 1.   Key Manufacturing Industries

A review of this basic information demonstrates the important role chemistry plays in our entire economy. Many also recognize that, not only are the direct contributions quite substantial, but indirectly, chemistry serves as an enabling foundation for most other manufacturing industries and its impact is actually enhanced as chemical technology works its way through petroleum, paper, electronics, automotive, materials, aerospace and biotechnology industries.(2) Unfortunately, at times, the positive contributions and key role of the chemical industry are not widely recognized.

The principal issue of concern is the economic value of the chemical industry and how continued investment in the chemical research infrastructure is critical to its health. The subject has been extensively evaluated, debated and assessed in professional society meetings and publications.(3-9) Key to these discussions is the longer term outlook for the chemical industry in anticipation of predicted modest or slow economic growth, saturated domestic markets and an increased focus on foreign markets. These factors, along with the cost and returns of conducting industrial research in the U.S., will continue to influence the future direction and location of research. Furthermore, in assessing the chemical industry, leading consulting firms are predicting further consolidation well into the 21st Century. This consolidation is similar to that which we have recently seen with pharmaceuticals and banking, and which is just starting for utilities. This anticipated consolidation activity, designed to improve productivity and efficiency could be a factor influencing the future of industrial chemical research.

In the next sections, we will examine the current status of research and determine the emerging factors that are influencing the industry and, in turn, their impact on the future of research. At this point, let's start by assessing the key industrial drivers that are affecting the profile of industrial research and then set forth the case for increased dependency on the knowledge integration process involving all facets of the research enterprise.

Industrial Drivers for Change in the 80's & 90's

Insight into the current and future trends for the research enterprise can be obtained by examining four industrial drivers that have influenced our industry. The most significant is globalization of markets and manufacturing; followed by the impact of the equity markets and business trends; also important is the role of risk, liability and regulation; and, finally, we will consider the perceived future value of chemical technology.

Globalization. On a relative basis, the chemical industry is not a pioneer in globalization. The petroleum and mining industries have led in this activity well before WW II. This trend, of course, is expected as a result of the worldwide distribution of natural resources. Conversely, motivation for the apparel industry in globalization is associated with the labor-intensive nature of product assembly and production. The electronic consumer products industry became globalized as a result of a rapid implementation of new technology and manufacturing efficiencies off shore, particularly in Japan and the Pacific Rim. The automobile industry is another major example of globalization by seeking large markets and efficient production.

As a result of global integration by so many of its customers, the chemical industry has been rapidly progressing to a fully worldwide integration of technology and markets. These trends began in earnest in the 70's and accelerated at a remarkable pace by serving customers globally and focusing on market needs. The most impressive example to demonstrate the extent of globalization of the chemical industry was recently presented.(*10*) Of the 1 million jobs provided by the chemical industry in the United States, almost one half of these are with global corporations having offshore ownership. This is an extraordinary transformation within the chemical industry in the last 20 years. Names such as Bayer, Hoechst, Rhone Poulenc, ICI, Elf Atochem, Akzo Nobel, Henkel, Mitsubishi, Nissan, BASF are highly visible components of the U. S. chemical industry enterprise as well as the U. S. equity markets. Each of these multinational global corporations is proud of its American citizenship. For the most part, they manufacture in the U.S. the overwhelming majority of products sold in the U.S. and commonly serve as net exporters of chemicals. Conversely, many leading American corporations are now reporting off shore sales and profits approaching and occasionally even surpassing domestic levels.

Globalization of industries has had an extremely profound effect on marketing, manufacturing and research and development. C. K. Prahalad articulates that as a result of globalization, there is no significant monopoly, nor any profit sanctuary for any company.(*11*) This phenomenon is most obvious to the average American consumer in the automobile industry. Until recent times, the "big 3", Chrysler, Ford and General Motors, dominated the U. S. automotive landscape. In the last 20 years or so, a highly competitive community has emerged, consisting of the "big 12 or more". Likewise, the chemical industry has undergone a similar transition. In earlier years, service-intensive specialty products enjoyed handsome margins and competition from two or less additional suppliers. Today's globalized marketplace has nearly eliminated these exclusive profit sanctuaries. For many significant markets, the number of chemical suppliers has more than doubled. Increased competition has strongly influenced the way corporations operate; consequently, the goals for marketing, manufacturing and research have given rise to: 1) an internal focus on market return, 2) timing that is now, not in the future, 3) a focus on the bottom line and on current business, and 4) an increased reluctance to bet on new and uncertain outcomes.(*11*) Thus, manufacturing, together with research and development personnel, have joined marketing teams in working closely with customers to address needs and to speed up delivery of new products.

Equity Markets - Business Trends. During the 1990's, there has been an emphasis on enhancing shareholder's value which, in fact, has become a dominant management philosophy in many circles. The influence of the equity markets has placed a strong demand on sustained quarterly performance, as well as market leadership. Industry has responded by establishing entrepreneurship and decentralization with individual empowerment by invoking the business unit concept. These factors have had further impact on research and development by permitting R & D to be more fully integrated into business planning and strategy. The reaction of the R & D community has been responsive, but concerned, as research horizons are reduced.(*12*)

Risk, Liability and Regulations. Environmental, health and safety issues have been a top priority of the chemical industry. The industry continues to make a major commitment to provide safe products based on best applicable technology and processes that meet or exceed standards.

Many scientists are gainfully employed developing new or improved product and process technologies to meet demanding safety and environmental goals. Some believe that this fruitful and personally rewarding activity has also affected the portfolio of industrial research programs by reducing the investment in future product development.

Technology Classification. Is chemistry a mature or an enabling technology? The debate has been quite extensive and eloquently championed by many, particularly Prof. Ronald Breslow(*13*) and the other authors of this publication. Overall, the concept of "maturity" influences the confidence of business managers on the role and impact of research and development investments. Maturity of markets and technology gives rise to a maintenance philosophy, where investment for growth becomes less common.

Chemical Industry Research - A Profile

In the last 20 years, changes in the economic community have exercised a profound impact on chemical industry research. Research organizations have undergone extensive review and evaluation in an effort to improve productivity, flexibility and responsiveness to customer-driven needs. Improvement goals in each of these categories have been set at 30 - 40%. To attain these goals, increased attention is given to aspects of core competencies, consolidation, organization, metrics, quality, etc. Consequently, central research has been downsized and applied R & D has moved to business units to focus resources and accelerate commercialization.(*14*) At the same time, the U. S. chemical industry's investment in R & D has increased only modestly, at a rate of about 3% in current dollars. Data for 23 leading chemical companies show an increase from $5.0 billion in 1986 to $6.7 billion in 1996,(*15,16*) see Figure 2. The overall employment in the chemical industry has been essentially constant from 1980 to 1996, at about 1 million. According to the Chemical Manufacturer's Association, this total includes 90,000 research and development scientists and engineers.(*1*) These investments for knowledge and technology development are being made in a highly competitive marketplace. Continued investment in industrial R & D has enhanced the ability and efficiency for conducting research with the application of advanced technology for computational molecular design, analytical instrumentation, refined synthesis capabilities, process automation, electronic communications, etc.

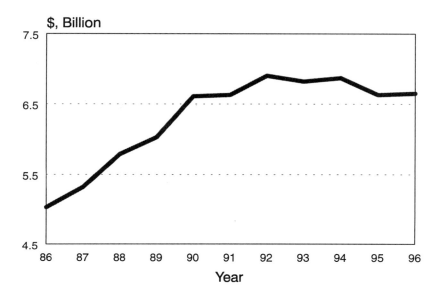

Figure 2. R&D Spending - Large U.S. Chemical Companies

One growing issue associated with the nature of chemical industry research is how it will change in the future. To adequately address this issue, it is necessary to define or classify research by type or category. Classical definitions of basic and applied research do not adequately reflect the close relationship and inter-dependency of research and development activities required for innovation. Today, many believe that long standing definitions on the value of basic vs. applied research are part of an outdated model of technology innovation.(*17*) To address this critical issue, while side-stepping the debate on defining various stages of research, we will employ information provided by the Department of Commerce which categorized U. S. industrial research of $110 billion in 1996 into the following facets:(*10*)

5%	Exploratory Research
20%	Applying knowledge to targeted needs
75%	Product and Process Development, Technical Service, etc.

This distribution reflects relative activities for industrial chemical research. The overall nature of innovation in industrial research and development is presented in Figure 3, which was adapted from the research work of R. Cooper and reports from the chemical industry.(*18*)

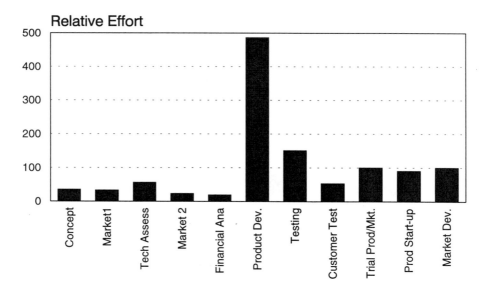

Figure 3. Industrial Research Profile, New Product Development

This plot of relative effort shows the center of gravity for industrial R & D allocations but does not include required capital investments. Nonetheless, this profile is quite instructive. It demonstrates that initial concepts constitute an important yet modest investment in the complex and uncertain process of innovation. Indeed, the initial step is critical and essential; however, a strong scientific/technology base is required for each of the following stages, bringing knowledge to bear on all of the many ensuing challenges. This process of internal knowledge generation and integration is essential for the successful implementation of new technology. Thus, during the recent past, industrial research has been focusing on those processes in Figure 3, which are more readily measured and accountable using accepted financial practices.(19) Consequently, industry will continue to be dependent upon the national research infrastructure to foster new knowledge for targeted areas and to conduct exploratory research that can enhance the development of initial concepts, which are the starting point of industrial innovation.

Globalization of R & D

In constructing this picture of trends and the future for chemical industry research, economists constantly remind us that "globalization" is serving as the most influential factor affecting business and markets and consequently, influencing research and development.

The U. S. has been a prime beneficiary of global research. The need to serve our attractive markets locally and gain access to the world leading education - research institutions, has attracted many new research facilities. Companies establish R & D facilities in different countries to a much greater extent than in the past. Initially, the driver for this investment was to address market needs but, more recently, the goals include engaging local talent to develop new products and processes and to keep pace with emerging technology.

The investment in research and development within the U. S. by foreign owned multi-national corporations is quite impressive and serves as a strong vote of confidence in U.S. markets and in the educational - research infrastructure. Between 1985 and 1991, foreign investments grew from 30% to 39% of the total chemical industry research in the U.S.(20) If this trend continues in the near future, research conducted in the U. S. by the chemical industry may be funded equally by national and off-shore companies. Conversely, approximately 18% of the total research conducted by the U.S. chemical and pharmaceutical industries is conducted off shore. Previously, research and development laboratories that were located outside the "home country" of a corporation, were given limited local responsibilities. As the laboratories grew and electronic communication technology evolved, a global corporate network has emerged. Corporations distribute research tasks globally based on a multitude of factors of efficiency, skills, capabilities, etc. Many research programs now have concurrent activities in several countries.(21)

Communication technology allows the deployment of resources for research and development almost independent of location. It rewards organizational openness. Networks leverage free minds and knowledge integration. Many chemical companies have informally announced successful innovations where coordinated R & D programs were conducted over several different laboratories, one within the U.S. and others off shore, e.g., Dow, Shell, ICI, etc. While there have been several worldwide conferences on managing a global research network(22), successful case studies and publications are relatively limited at this time. Exxon attributes the new lubricant development EXXSYN 6 to an integrated R & D effort at Baton Rouge, LA; Sarnia, Ontario; and Abingdon, England. 3M similarly reports new adhesives research in Neuss, Germany and St. Paul, MN., abrasives research in Atherstone, England and St. Paul and optic media research at 3M Japan and St. Paul.(23)

Akzo Nobel, a multi-national health care, chemicals, coatings and fiber corporation, operates 35 research laboratories in 10 countries. The central, chemicals research organization conducts higher-risk, exploratory research at three locations, Germany, the Netherlands, and the U.S. Global cooperation for new product development is becoming more commonplace. As an example, a new water-borne coating system is being developed in Duran, England, Arnhem, the Netherlands and Dobbs Ferry, NY. New intermediates process technology is being developed in Dobbs Ferry, NY and Obernburg, Germany, for eventual implementation in West Virginia. Similarly, Nestle's operates 17 R & D centers in 11 countries. ICI coordinates synthesis and development for new agricultural products at laboratories in England and in Richmond, CA. The trend continues to grow.

The integrated management of global R & D will continue to gain prominence as corporations must establish a presence in an increasing number of locations to access new knowledge and to absorb new research results from universities and competitors. As stated earlier, global competition is forcing companies to move new products from the idea phase to the marketplace ever more rapidly. Consequently, R & D networks will be designed to tap new centers of knowledge as these emerge across the globe and to accelerate commercialization of products in foreign markets. (24)

Current and Future Trends for Industrial Research

As we approach the 21st Century, chemical research and development is beginning to reclaim its position as a vital resource and an essential investment for the continual renewal of the U. S. chemical industry as the demands from the global economy intensify. During the last fiscal year for which data is available, 1996, 10% of the sales of the chemical industry were attributed to new products and new processes.(25) In addition to this solid performance, there are many emerging technologies attesting to the value of exploratory/discovery R & D in a time of increasing competitiveness. A representative sampling (non-comprehensive) of the innovative vitality of the chemical industry is given below:

Innovative Developments

1.   Metallocene catalysts innovations for polyolefins copolymers - Dow, Exxon, Phillips Petroleum, etc. - A host of new specialty performance products prepared from available monomers

2.   Catalysts and polymer developments for polyolefin ketones - Shell

3.   Polymers of dicyclopentadiene from metathesis catalysis – Phillips Petroleum, Hercules (Meton), Goodrich (Telene)

4.   Catalyst technology for a new composition polyester carpet yarn - Shell

5.   Catalyst system for ethylene copolymers U. of N. Carolina/DuPont

6.   Special zeolite catalysts for selective manufacture for functional chemicals, e.g., butyl amines - BASF, pharmaceutical intermediates - several companies.

7.   Selective oxidation of olefins to manufacture specialty intermediates – Eastman Chemical

8.   New shape selective catalyst for petroleum refining - Exxon, Mobil, Chevron, Caltech/Akzo Nobel

9.   Selective catalysis/process technology, butadiene to maleic anhydride - Dupont

10.   Process & catalysts technology to convert natural gas to valuable liquid fuel - Exxon.

11.   Selective oxidation - benzene to phenol, Monsanto/Novosibirsk

12.   Anti-cancer candidates with complex large structures are easily synthesized with new metathesis catalyst - example of where research on polymer chemistry has helped organic synthesis

13.   New sweetener, neotame, 8000X sweeter than sugar - Monsanto

14.   Olestra - non-digestible fat - Procter & Gamble

15.   Stereoselective catalysis - pharmaceutical applications – several

16.   Chlorofluorocarbon (CFC) replacements - refrigerants, solvents - DuPont

17.   Fused glass composite single filament fiber - Microflex - Owens Corning

18.   Selective oligomerization of ethylene to 1–hexene - Phillips Petroleum

19.   Nano-dispersed Nafion on silica - acid catalysts - DuPont

This brief list of examples serves not only to demonstrate the vitality of industrial research but also it gives evidence of the more recent effort to balance research portfolios with new discovery R & D programs. A recent description of central research strategy at DuPont and Exxon serves to further demonstrate the investment value of a balanced portfolio.(26)

Traditionally, corporations have responded to competitive pressures by investing in people, training, technology, and better products. It is expected that this approach will continue to dominate the U.S. landscape as our nation will need to maintain high technology positions as current manufacturing technology becomes globally dispersed.

Incremental Development.   Over 90+% of current industrial research activities satisfy the key goals associated with metrics, quality, timetables, and return on investments. These programs fulfill a critical task of making research vital to the corporation. A significant component of this activity is the continuous improvements that are achieved via novel or modified technology for existing processes and products. Unfortunately, this process is frequently referred to by the uncomplimentary term of incremental development. This terminology does not adequately reflect the resources, skills, personal commitment, science and technology required for a successful program. Actually, the results can be rather spectacular. Shell announced that the cumulative results of continuous evolution and incremental improvements for manufacture of ethylene oxide are stunning. Accordingly, after 20 years of continuous improvements via process engineering technology, catalyst technology, etc., the advances in selectivity, yield, productivity and energy efficiency have provided an overall economic improvement far greater than the original process. This type of result would normally be expected from only a major step change with a totally new breakthrough technology. Thus, this important procedure of continually bringing new engineering and science to current products and processes will continue to be essential for the technologically demanding and highly competitive global market. These are significant and valuable accomplishments.

The responsibilities of industrial R & D chemists and engineers are expanding widely. They are expected to work effectively in teams, to provide the skills and capabilities for interacting with customers, to work with manufacturing, to address health and safety services, and to deal with legal or intellectual property issues, while keeping updated on emerging technologies. This process is placing greater entrepreneurial responsibility on each member of the research organization. In addition, these R & D responsibilities will continue to provide opportunities for professional growth and to offer educational challenges for academia.

Partnerships.   Since the mid-70's, central industrial laboratories have been shrinking to increase efficiency, focus on core business and to foster flexible entrepreneurship. As a result, more resources are going to product development and technical services activities, because it is essential to maintain close ties to your customers in today's highly competitive marketplace. Emerging from this transition is a new role for corporate R & D: to facilitate a network of activities involving cooperative interaction among the dispersed global laboratories, and in particular, "to expanding beyond the box" to involve external sources, such as universities, government laboratories, consortia and even competitors.(27)   While central R & D has been diminished, it is fulfilling an influential role of helping to develop and coordinate strategic technology issues on a global scale.(28)

The traditional philosophy of a self-sufficient, internal chemical research appears less viable now with a globalized economy, than in the past. No one company can readily master each aspect of technology required to maintain market leadership on a global scale.(29) External partnerships: industry - industry or industry - academia - government provide many mutually beneficial advantages. There has become a ground swell of interest to fulfill needs for discovery research in a time of great fiscal constraints.(30) Advantages for research partnerships are listed in Table 1 and the growth of these associations is shown in Figure 4. Leading the list is the access to new knowledge and concepts being generated in universities, institutions and national laboratories.

Together with the insight on emerging technologies and different perspectives, one can assess technology options with greater efficiency. The process of technology assessment is accelerated by focusing effective teams with leading experts. This process provides greater R & D flexibility by utilizing skills not readily available internally without the need, in the initial phases of the research program, for adding the specific skills. By leveraging resources with other partners, one can take on higher risk discovery research programs in today's climate of intense competitiveness for internal resources. These longer term partnerships are quite beneficial to an organization with a charter to help establish technology strategy and to enhance recruiting. Furthermore, many industrial scientists enjoy the opportunity of participating in a fundamental, exploratory research program. They find this activity quite motivating as it balances the experience of internal programs addressing more immediate needs. This activity of interacting with leading experts to establish strategy and technology assessment serves as an educational, renewing process for a research organization.

Table 1. Research Partnerships - Advantages

- New knowledge and concepts

- Insight on emerging technologies

- Different perspectives

- Accumulated technology assessment

- Greater R&D flexibility

- Leverage resources & reduce risk

- Justify higher risk – exploratory research

- Long term relationships

    - consulting – strategy - recruiting

- Motivating scientists

- Renewing the organization

For more than 20 years, the National Science Foundation has pioneered efforts in promoting partnerships by establishing centers of excellence and cooperative research consortia.(*31*) These consortia fulfill an important role in knowledge integration by bringing together a multi-disciplinary set of skills to address a target technology and permit any corporations to participate and contribute. The recent trend illustrated in Figure 4 demonstrates a desire by industry to interact more closely with individual professors to achieve specific research goals.(*29*) This approach to research through centers of excellence and individual contract programs provides a cost effective alternative to the more traditional method of conducting discovery research.

The R & D organization chart shown in Figure 5 illustrates the principal partnership interactions for a central research organization. The core enabling technology teams are responsible for the discovery research on emerging technologies, principally with national laboratories and universities, and, subsequently, for transferring these internally refined technology leads to a business focus R & D group. This group maintains the major responsibility for teaming with customers, suppliers and competitors.

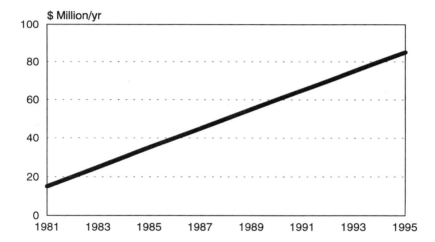

Figure 4. Academic Industry Alliances, Chemistry Departments

The chemical industry is somewhat behind other industries, such as electronics, in cooperative R & D development with competitors. Of course, we have a history of establishing formal Joint Venture business to bring together technologies, manufacturing and markets, but these less formal cooperative R & D programs are beginning to grow in number. One recent example in this area is the Chemical Industry Environmental Technology Projects. A consortia of eight companies, DuPont, Air Products, Akzo Nobel, Oxychem, NL Industries, Praxair, FMC, and Battelle, are working jointly on R & D programs associated with environmental remediation, pollution prevention, biological treatment systems, etc. In a specific project on HC1 recovery for conversion to chlorine, approximately 15 companies are cooperating in a development project.(*32*)

Company — company partnerships are certainly not new to our industry but are becoming more common place for all the reasons listed in Table 1. These associations are especially driven by the cost of R & D and the opportunity to accelerate development. Currently, Dow and DuPont are reportedly cooperating on an elastomers technology program that probably would not have occurred with the independent philosophy of just a few years ago. The trend is gaining acceptability where once there was substantial resistance. Recently, in discussing the new engineering plastic based on polyolefin ketones, a Shell representative stated that "if it had to do it all over again today," they would probably seek several cooperative partners rather than conduct a solo R & D act as was done.

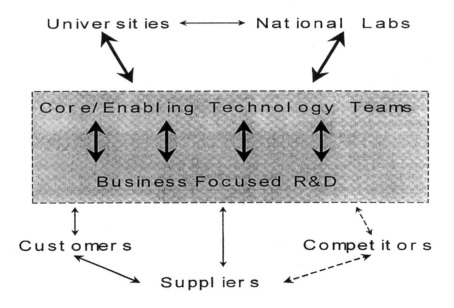

Figure 5.   R&D Organization - Knowledge Integration

Many of the leaders in the chemical industry are also leaders in the arena of technology acquisition through external partnerships. Full time positions or entire departments have been established to address this opportunity. In the last five years, DuPont has nearly tripled its resources for external partnerships from about $15 million/yr. in 1992 to about $50 million in 1997. In this effort, nearly 2,000 ideas are reviewed each year for programs with the universities and national laboratories. A favorite success story on partnerships for DuPont is associated with their new refrigerant business. DuPont made the switch from Freon chemicals to a new line of hydrochlorofluorocarbons, sold under the tradename Suva, with remarkable speed. The new products, with $480M in new plants, were on line in only four years, where, typically, a 7 year cycle is the norm for transition from concept to manufacturing at this magnitude . This is quite a remarkable achievement for the chemical industry. The accelerated research and development is attributed to collaborative efforts which included two government agencies, the National Institute of Standards and Technology (NIST) and the Department of Energy (DOE). The team consisted of approximately 13 national labs and 10 universities. The task involved screening hundreds of candidate compounds, selecting viable candidates and developing manufacturing processes.(33) Interestingly, chemistry and chemical engineering departments conducted part of the task of identifying the best compounds and developing flow sheets for manufacturing.

Additionally, DuPont reports successful partnerships with the universities and national laboratories for accelerating programs and bringing new concepts on line in areas of catalysis, polymer technology, agricultural and medical products. Many of the chemical companies on the Fortune 200 list also have reported commitments and successes in partnerships with universities and national laboratories. These include Chevron, Gillette, Kodak, Eastman Chemical, Dow, Lord Corporation, ICI, Monsanto, Procter & Gamble, Air Products, etc.(34) Our company, Akzo Nobel, has published the results of successful partnerships in polymerization catalysis, petroleum refining catalysts, chemical synthesis via zeolites, corrosion resistant coatings and copolymers for water borne-coatings.(29)

In one of my favorite examples of an external partnership, we assembled a team to evaluate an internally developed concept, by Dr. M. Geerlings, for cancer treatment with alpha radioimmunotherapy. In this case, the question is asked: "Can alpha radiation treatment prove advantageous over the current beta radiation treatment?" The task is shown in Figure 6.

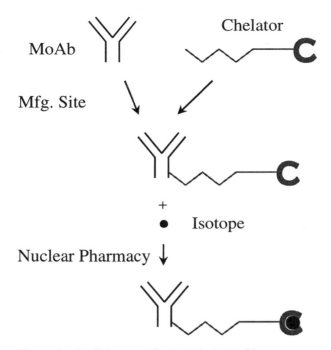

Figure 6.   Radioimmunotherapeutic Assembly

An appropriate radioactive isotope is bound to a monoclonal antibody that is sensitive to acute myelogenous leukemia and selectively carries the radiation source to the target site. The task is complex and requires a multitude of disciplines to demonstrate feasibility. The team consisted of three funding sources: Akzo Nobel - initial concept and characterization chemistry, Protein Design Labs - source of the monoclonal antibodies and NIH - funding for pretesting and preclinical trials. Four national labs participated by providing the nuclear chemistry technology. Oak Ridge supplied the precursor to the isotope of choice Bi^{213} which is a reportedly "patient friendly" source of alpha radiation. The Karlsruhe laboratory in Germany purified the precursor and developed a delivery system. Battelle Pacific Northwest determined the dosage rate by computational modeling and the Los Alamos laboratory, along with the National Cancer Institute, synthesized several chelating agents necessary to hold the active isotope as the assembly swims through the body. This binding of the isotope to the antibody is essential. If the isotope is pulled from the monoclonal antibody assembly by any one of the multitude of chelating agents within the human body, the approach is ineffective. Finally, preclinical testing and limited clinical trials were conducted at the Sloan Kettering Center in New York City. Preliminary results were reported(*35*), indicating that this novel approach is 10,000 times more selective than the currently used I^{131}, a beta emitter. Full clinical trials are planned. Thus, in a relatively short time, we were able to accomplish a complex task by utilizing external sources and capabilities. This example also serves to illustrate the enabling contributions of chemistry at the interface of biotechnology.

The resources available for partnerships in North America are extraordinary and quite diverse. In the fields of chemistry, chemical engineering and biotechnology, there are more than 53,000 professors, post doctoral candidates and graduate students at North American universities. Also, there are more than 35,000 scientists at the top 14 DOE national laboratories,. Add to this the thousands of scientists at NIST, NASA and Department of Agriculture laboratories and one can begin to recognize that within this extraordinarily vast national resource, there must be scientists and engineers that can contribute to any particular knowledge and technology area.

Many of our academic research colleagues report that these cooperative programs permit scientifically interesting research on challenges that could have tangible, practical impact. The interdisciplinary nature of these alliances expand traditional boundaries and, thus, serves to extend the researcher's horizon. Team members indicate that the opportunity to participate in the "front" part of our industrial research project is a professionally beneficial experience. Students have found that these experiences are helpful in initiating a career in industrial research.

These partnerships are not substitutes for internal programs but provide an expansion of the research horizon for higher risk, discovery R & D. There are many government agencies at the Federal and State level that actively participate in R & D partnerships for economic development. We have many examples in New York, Virginia, Pennsylvania, etc., working principally to fulfill a vital incubator role. Leading chemical companies have been involved with DOE and DOD, while these are important activities, they represent a very small portion of chemical industry research(*10*), reportedly less than 3%.

Global Partnerships. Multinational corporations are seeking knowledge and cooperative associations on a global basis. Most leading U. S. chemical companies, with active external technology programs, have assigned managers, on a full time basis, to seek and establish cooperative programs in key geographical centers - Europe, Japan, Russia, China, etc. U.S. companies have expanded external activities and participate in government co-sponsored programs in Europe. A major characteristic of R & D funding in France, Germany, Belgium, the Netherlands, etc., is the existence of a variety of government funded research institutes that bridge the research and development activities of industry and universities. This activity has been well documented.(*36*) Thus, I would like to review a recent activity in research globalization: "The Rush to Russia." The end of the cold war and the collapse of the former Soviet Union have created challenge with social - economic - security - technology issues for many nations, especially the United States. This change suddenly opened a vast pool of technologies and skilled talent to the reservoir of accessible scientific resources. At the encouragement of our national government, many corporations have come forth to explore opportunities for cooperative research.

While publication of these activities are limited, a few examples can serve to illustrate the strong interest in this technology resource. Like many multinational chemical companies, Akzo Nobel is funding several individual programs on phosphorus chemistry, polymer chemistry, and catalysis in Russia. In addition, a visiting professor from Belarusian University in Minsk, is on a sabbatical at our Dobbs Ferry, NY research center. Others are far more active. DuPont has established more than 50 research programs within Russia. At least 10 major petroleum and chemical companies have established research programs that range in size from 2 to 100 participants at the Novosibirsk catalysts research center. In fact, the recent development announced by Monsanto for direct conversion of benzene to phenol is attributed to a cooperative program at Novosibirsk. With a more aggressive plan, Corning has contracted an entire Materials Research Institute in St. Petersburg. After several years, further investments were made for a separate research facility and more than 2 dozen of these scientists were hired as full time Corning employees.(*37*)

During the next few years, we will be learning much more about the potential value of this venture. Nonetheless, the concept of research partnerships is expanding to all corners of the globe.

The Future Team

As a result of this teaming exercise over the last 10 years, more industry, academia and government scientists are generating a stronger curiosity about the needs and interests of each other. The lead players recognize that the strength of a true team is far greater than the sum of the individual components and have begun to move in concert. Industry is beginning to learn to work more openly and has enhanced its commitments to funding and to participating in cooperative programs with other corporations, universities and national laboratories. Universities, with their strength in exploratory and discovery research, are becoming more entrepreneurial and recognize the benefit from teaming with industry. The most critical steps in this direction are being made by the 3rd team member, the government.

NSF. The National Science Foundation has laid the groundwork for many years of funding individual investigators and centers of excellence such that fundamental knowledge and discovery research can serve as a basis for strong partnerships with industry. Recent action in establishing goals for knowledge integration, in increasing reliance on partnerships and in promoting personnel exchange programs under "GOALI" is fostering industry university collaborations, while maintaining a major commitment to the individual investigator.

The National Laboratories

The establishment of the cooperative research and development agreements (CRADA), by NASA and, particularly, by the Department of Energy have proven most helpful to industry by permitting access to world leading scientists and sophisticated capabilities on an equal, co-funding basis. In many cases, these activities are mutually beneficial. Concerns are mounting as to the commitment of government to these partnerships.(*38*)   The roller coaster ride for DOE CRADA's is shown in Figure 7.

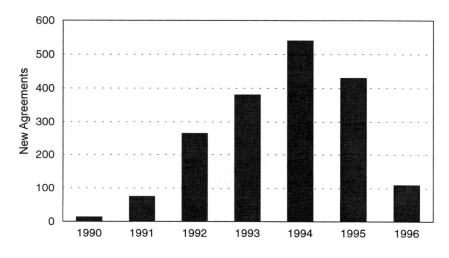

Figure 7.   A Roller Coaster Ride for CRADAs

Part of this surge and retreat has arisen as the national labs seek to define or expand missions as a result of the end of the cold war and increased economic competition. Two recent studies have been made.(*39*) Industry members associated with external research programs, including CRADA's, found the testimony of Dr. Sattelberger to the House Science Committee to reflect their experience.(*40*)

Defining compelling missions, while important, should not restrict the options of the laboratories. We should encourage them to use their tremendous capacity for interdisciplinary R & D to push the limits of science and technology in the national interest.

Measuring the success of CRADA's has been rather difficult, given their brief history. Nonetheless, our experience and many of my colleagues have indicated that the knowledge exchange has proven mutually beneficial.

NIST. Another important government action in partnerships has come from the National Institute of Standards & Technology (NIST) with the Advanced Technology Programs. The U. S. government with this agency has taken a very bold step of funding, jointly, with industry, knowledge integration research which ventures far along the innovation chain, shown earlier in Figure 3. This adventure has attracted substantial controversy, but we are encouraged that a continued commitment is allowing this program to consider two focus areas for the chemical industry: catalysis and separation.

An Expanded Role. We know that there are many benefits from the integration of industry, university and government research skills and capabilities. As we march into the 21st Century, government's role for chemical technology is becoming more critical,(41) see Table 2.

Table 2: New Chemical Science & Emerging Technology

Government's Role for Chemical Technology

1. Partnerships required in both science and technology

2. Exploratory/discovery research in jointly selected target areas

3. Long term, high risk, broad applications, new knowledge - shared expense

4. Enhance the research infrastructure to continue to get R & D capital spent in the U.S.

5. Balanced portfolio of programs in areas of national needs

The strategy outlined in Table 2 is a modest extension of current practices. Also, this approach is commonly practiced in the Netherlands, France, and Germany and Belgium where industry plays a key role in planning and identifying key technology issues. Most recently, the United Kingdom has re-examined its approach to government sponsored research and has conducted an assessment for the Technology Foresight Program.

Technology Foresight - United Kingdom. The UK Cabinet Office of Science and Technology established the Technology Foresight Program as a multi-year comprehensive study to help identify emerging opportunities in markets and technology and assess areas of strength and areas of need for the UK research infrastructure. The study consisted of 15 sectors, including chemicals, energy, materials, agriculture, health and life sciences, electronics, etc. Interestingly, chemistry was reported in Vol. I. The key conclusions of the study were given in five areas - catalysis, chemical engineering, sensors and measuring, materials polymers and biochemical technology. I have followed the activities in catalysis, and can report that as a result of the Foresight study, industry and government are jointly funding a virtual center of applied catalysis. A full time director has been appointed and several universities and industry laboratories are participating in the initial programs.(*42*) If the center develops as anticipated, it should provide opportunities for many global chemical companies to conduct cooperative research programs.

Vision 2020 - The U. S. Chemical Industry. At the request of the White House, Office of Science & Technology Policy, and the Office of Industrial Technology of DOE, representatives of the chemical industry produced a road map for the chemical industry.(*43*) This document, Vision 2020, prepared under the leadership of John Oleson, Dow Corning, provided advice on how the U. S. Government could allocate R & D spending to advance the manufacturing base of the U. S. economy.(*44*) One part of this program is associated with providing a vision for new chemical sciences and engineering technology. Under the leadership of Victoria Haynes, B. F. Goodrich, key technologies were identified and committees were established to set forth specific recommendations in each technology issue. A brief outline of the technology topics is provided in Table 3.

Table 3: New Chemical Science & Emerging Technology

Chemical Synthesis & Catalysis
Bioprocessing and Biotechnology
Materials Technology
Process Science & Engineering Technology
Chemical Measurement
Computational Technologies

For each of the prime categories listed in Table 3, a detailed consensus is being developed with the scientific and industrial community.(*45*) As a result of my interest in catalysis, we will concentrate on this one area, as an example of the Vision 2020 process.   In March 1997, a representative team of about 55 technology leaders held a workshop in Washington, D.C. (*46*)   An outline of the conclusions, for only the segment on research targets, of this two day workshop is given in Table 4.

Table 4: Catalyst Technology Roadmap

High Impact Areas for Catalysts Research

>   Selective Oxidation
>   Alkane Activation
>   Green Process Technology
>   Stereoselective Synthesis
>   Functional Olefin Polymerization
>   Alkylation
>   Living Polymerization
>   Alternative, Renewable Feedstocks

The process of identifying key technologies and focusing on specific research goals is not without controversy. In fact, during our initiating meetings on the Vision 2020 technology issues, the question was raised "How do you put your arms around an elephant?" This comment is in reference to the vast diversity of technology issues for the chemical industry. The key point is that this process has begun and through a series of refinement activities with the involvement of the technical community, a general consensus will be developed to assist the partnership process.

As in the past, industry's growth will be highly dependent upon the outstanding talent pool graduated from the education/research infrastructure. For all the factors associated with technology evolution and economic globalization, this dependency is growing and is fostering a cooperative spirit including the government and academic sectors together with industry to lead us in the 21st Century. The goal is to establish a balance of curiosity-driven and goal-oriented, fundamental and discovery research. Government has a vested interest in basic, discovery research and must share expenses in high risk projects which have a big payoff .(*41*) In the next century, many of us and our successors, working jointly with each segment of the research enterprise, will be expanding the frontier of chemistry and playing an enabling role at the interface with many other vital technologies: e.g., health, energy, materials, electronics, environment and computational science.

Acknowledgements

I am thankful for the many discussions and suggestions of my colleagues, Ed Walsh, Dave Marr and George Whitwell; for input from the information scientists, Ed Santos and Ray Davison; for Jacqueline Lacerra, who delayed retirement by several weeks to type and re-type this paper; and for my family for their patience and understanding.

Literature Cited

1. Lenz, Alan J., Director, Economic Analysis, Chemical Manufacturer's Association, presentation at American Chemical Society National Meeting, 9/8/97, "What is the Future of R&D in the U.S. Chemical Industry". Thayer, A.M., "Industrial R & D Scrutinized", *Chemical & Engineering News*, Sept. 22, 1997, pp 14-15.
2. Critical Technologies: The Role of Chemistry and Chemical Engineering, National Research Council, Washington, D.C., 1992.
3. Myers, M. B. & Rosenbloom, R.S. "Rethinking the Role of R & D, *Research Technology Management,* **1996**, *39*, pp 14-18.
4. "What is the Future of R & D in the U.S. Chemical Industry", Am. Chem. Soc. National Meeting, 9/8/97.
5. Samid, G., "National Funding Strategies for R & D" *Chemtech*, Sept. 1997, pp 7-11.
6. Workshop, "Global Research Management," Industrial Research Institute and MIT, Washington, D.C., 7/30/96.
7. Workshop, "Organizing for R & D in the 21st Century," NSF & DOE, Washington, D.C., April 24 –26, 1997.
8. Roussel, P., Saad, K., Erickson, T., *Third Generation R & D*, Harvard Business School Press and Arthur D. Little, Inc., 1991.
9. Annual Meetings of the Council for Chemical Research.
10. Good, M. L., "The Federal Role in Technology and Competitiveness: Impact on the Chemical Business". Unpublished communications.
11. Prahalad, C. K., *Competing for the Future*, Harvard Business School Press, 1994. Prahalad, C. K., *Research Technology Management*, Dec. 93, pp 40-47.
12. "R & D Spending Retreats Once More"", *Business Week*, June 27, 1994, p 78.
13. Breslow, R. *Chemistry Today and Tomorrow*, American Chemical Society and Jones and Bartlett Publishers: Washington, D.C., Sudbury, MA, 1997.
14. *Current Trends in Chemical Technology, Business to Employment*, American Chemical Society, Washington, D.C., 1994.
15. *Chemical & Engineering News*, Sept. 1, 1997 p 52.
16. Science and Engineering Industries – 1996 (NSF), p 113.
17. Myers, M.G., Rosenbloom, R.S., "Rethinking the Role of Research", *Research Technology Management*, June 1996, pp 14-18.
18. Cooper, R. G. and Kleinschmidt, E. J. *Research Technology Management*, August 1996, pp 18-29. Cooper, R. G. and Kleinschmidt, E. J., *Product Innovation Management,* **1995**, *12*, pp 374-391, and reference cited.

19. "Integrating Technology and Business Planning in IRI Companies", Industrial Research Institute, 202 – 296 – 8811, Washington, D.C. "The Research Value Pyrimide", A metric guide to R & D, Industrial Research Institute, Washington, D.C.

20. *Science & Engineering Indicators*, Nat'l Science Foundation – 1996.

21. Harris, C. R. C., Insinga, R. P., Morone, J. P., Werle, M. J., "The Virtual R & D Laboratory", *Research Technology Management*, April 1996, pp 32-36.

22. Workshop on Global Management of R & D, Industrial Research Institute, IRI – MIT 2 year study, Washington, D.C., 7/30/96.

23. *The Changing Global Role of R & D Functions*, The Conference Board, New York, NY, 1994. While 3M is not classified as a chemical company and its performance is not shown on any of the tables in this paper, we understand the important role of chemistry as a foundation to its many products.

24. Kuemmerle, W., "Building Effective R & D Capabilities Abroad," *Harvard Business Review*, April 1997, pp 61-70.

25. Industrial Research Institute/CIMS Annual R & D Survey, fy 92 – 96, Industrial Research Institute, Washington, D.C.

26. Miller, J. A., "Basic Research at DuPont", *Chemtech*, April 1997, pp 12-16. Eidt, Jr., C. M., Cohen, R. W., "Basic Research at Exxon", *Chemtech*, April 1997, pp 6-10.

27. Fusfeld, H. I., *Research Technology Management*, August 1995, pp 52-56.

28. Chiesa, V. *Research Technology Management*, October 1996, pp 19-25.

29. Via, F. A., *Chemtech*, March 1994, pp 10-18. Vleggaar, J. J., *Business Strategy*, April 1991, p 8. Via, F. A., ACS Symposium, Division of Petroleum Chem., 1992, 37, 1577.

30. Scientific Resource Committee of the Council for Chemical Research, 1994.

31. Illman, D. L., "NSF Celebrates 20 Yrs. of Industry – University Cooperative Research," *Chemical & Engineering News*, January 1994, pp 25-30.

32. Chemical Industry Environmental Technology Projects, LLC, Columbus, OH, 614-424-4724.

33. Guschl, R. J., *Chemtech*, July 1997, pp 7-10.

34. Council for Chemical Research Meetings or the External Research Director's Network of the Industrial Research Institute.

35. Scheinberg, D., Am. Assoc. Adv. Sci., National Meeting, Feb. 1997.

36. Abramson, H. N., Encartonacao, J., Reid, P. P. and Schmoch, U., "Technology Transfer in Germany", *Chemtech*, January 1998, pp 14-23. Effective Collaborative R & D, EIRMA WORKSHOP VI, European Industrial Research Management Association, 1996. Worldwide Industry – University Government Research Collaborations, Council for Chemical Research, Inc., Oct. 1995, Pittsburgh, PA Workshop. Reklaitis, G. V., *Chemtech*, May 1997, pp 10.

37. Thiel, F., Am. Assoc. Adv. Sci., National Meeting, Feb. 1998.

38. Lawler, A., "DOE to Industry: - So Long Partner", *Science*, **1996**, *274*, p. 24.

39. Reference 41. Galvin, R. W., Chair, Task Force Report: *Alternative Futures for the Department of Energy*, National Laboratories. Press, F., Chair, *Allocating Federal Funds for Science & Technology*, National Academy of Sciences.

40. Sattelberger, A., Director of Science & Technology Board at Los Alamos National Laboratory: Testifying to the House Science Committee, Feb. 28, 1996.

41. Miller, J., "R & D Cooperating and Funding vs Global Leadership". American Chemical Society Meeting, March 24, 1996.

42. Technology Foresight – Office of Science & Technology, Albany House, 84-86 Petty France, London, SW1H9ST, England, Director of Applied Research, Virtual Center, Dr. Christopher Adams, chris.adams@iac.org.uk

43. *Technology Vision 2020, the U.S. Chemical Industry*, The American Chemical Society, American Institute of Chemical Engineers, The Chemical Manufacturers Association, The Council for Chemical Research, and the Synthetic Organic Chemical Manufacturers Association; Washington, D.C.; 1996.

44. The chemical industry is represented by the American Chemical Society, the Chemical Manufacturer's Association, The American Institute of Chemical Engineers, The Council for Chemical Research, and the Synthetic Organic Chemical Manufacturer's Association. Leadership for the Steering Committee was provided by John Oleson, Dow Corning, and Leadership for New Chemical Sciences and Engineering Technology was provided by Victor Haynes, B.F. Goodrich.

45. The Council for Chemical Research, Vision 2020 Committee, Thomas Manuel, Air Products, Chair; Raymond Seltzer, Ciba, Past-Chair.

46. Workshop sponsored by Department of Energy and Council of Chemical Research. Workshop was led by N. Jackson of Sandia and T. Baker of Los Alamos.

11

Chemical Research and the R&E Tax Credit

Brandon H. Wiers

Manager, External Research Programs (Retired)
Procter & Gamble Co.
11261 Hanover Rd., Cincinnati, OH 45240

Federal tax provisions under the heading of the Research & Experimentation (R&E) Tax Credit establishing incentives for private sector investment in research are examined with a view to improving the support by corporations of research at universities and at government labs. Specific Tax Code amendment recommendations are made to (a) expand the definition of basic research, (b) establish the equivalency, for tax credit purposes, of government labs and universities, and (c) increase the credit rates both for general qualified research expenditures and for basic research.

Introduction

Public funding of federal agencies and government labs has come under increasing downward pressure in recent years. Because of this, but also because basic research costs have been rising steeply, research universities are also experiencing difficulty meeting their financial needs. At the same time, due to competitive pressures, research-intensive corporations have been reducing or eliminating altogether their basic research organizations. The livelihood of basic, long-range research thus appears seriously threatened.

Since it is clear that the private sector chemical research enterprise depends on universities for the training of their technical employees, the question is asked whether it shouldn't fall to the private sector to pick up the obligation to provide the needed funds for basic research. The issue is not entirely new, and in fact Congress some time ago provided an incentive to industry to invest in university basic research. The vehicle is the Research and Experimentation (R&E) Tax Credit, created in 1981 as a part of the Economic Recovery Act. At its inception it was given a one-year life. It has come up for renewal annually since then and been allowed to lapse on three occasions, only to be reinstated retroactively and then extended until June 30 of the coming year. Its current status is that it is in effect until June 30, 1998. Legislation is being considered at this time which would make the tax credit permanent.

The following discussion aims at showing how the credit might be improved and made to perform in a way that would tend to meet more effectively the needs of both the chemical research enterprise and the nation as a whole for more basic research investment.

What is the R&E Tax Credit?

Officially called the "Credit for Increasing Research Activities," it is one of a dozen so-called "General Business Credits" that a corporation may claim on its Federal Income Tax. Along with such other options as "Welfare-to-Work Credit," "Disabled Access Credit," "Credit for Alcohol Used as Fuel," and "Low-Income Housing Credit," it is intended to serve as an incentive to behaviors deemed beneficial to the nation's economic health and welfare (originally economic recovery). For guidance of those unfamiliar with its provisions, the language of Sec. 41 of the IRS Code has been expressed in equation form in Appendices A and B, together with worked examples.

Key to evaluating the credit's benefits are the following features.

- Corporate R&D expenditures are deductible expenses whether or not a credit is claimed. For a company in the 35% tax bracket, this means a savings of $35 for every $100 of R&D expenditure, i.e., a net cost of $65.

- Only a limited set (typically about 50%) of R&D expenditures actually qualify for consideration as the basis for a credit claim. For the sake of distinguishing these from the more general expenditures, they are referred to as R&E expenditures.

- The amount of the credit is currently required to be reckoned at a rate of 20% of the actual R&E expenditure, and the amount of the credit must be added back to the taxable income. What this means is that for every one dollar of R&E expense, a credit of $.20 is permitted and the same $.20 is added back to taxable income. Thus, for a company in the 35% tax bracket, the $.20 would result in an additional tax obligation of $.07. The net gain is therefore $.13 for every dollar of R&E expenditure. In other words, for every $100 of R&E expenditure, the savings due to the credit would be an additional $13 on top of the original savings of $35 due to deductibility, for a total savings of $48 for every $100 of R&E expenditure, i.e., a net cost of $52.

- There have been from the beginning two classes of R&E expenditures for which tax credit may be claimed. One class is called "Qualified Research Expenditures." The other is called "Basic (University) Research Payments." The former includes various in-house expenses, but also expenditures for outside (contract) research, only 65% of which are permitted to be included in the credit calculation. Basic (University) Research Payments, on the other hand, are treated as more credit-worthy than contract research by being allowed to enter the credit calculation at a full

100%. The base levels mandated by the law differ as between Qualified Research Expenditures and Basic (University) Research Payments (Appendix A); however, it is clear that the policy goal was to promote insofar as possible industry investment in university research.

• As originally written, the law required that there be a positive increment in R&E expenditures of both classes in order for a credit to be claimed. A 1997 modification to the law permits even those companies whose expenditures may have declined from one year to the next to claim a credit, provided those companies agree to continue to use the same accounting method henceforward. It is a three-tiered credit system that is covered, with an example, in Appendix B.

• Finally, the law as written omits specific reference to government labs. Consequently, the tax benefit implications to a company making research payments to a government lab are no different than doing business with a contract lab, for which expenditures are credited only to the aforementioned level of 65%. This is still a credit-worthy investment by a corporation, but clearly it has a different status than university research. (Other provisions favoring university research further underscore their preferred vendor status. See Appendix A.)

What has the R&E Tax Credit Accomplished?

Performance data are approximate and now somewhat dated, the most recent government assessment having been completed in 1995.(*1*) What the data indicate, nonetheless, is that of an industry R&D expenditure level of $83 Billion in 1992, a little more than half—$43 Billion—met the IRS requirements for qualified R&E expenditure. The resultant tax credit value of this amounted to an estimated $1.6 Billion.

As has been true since the beginning, the R&E Tax Credit claimants are predominantly manufacturing firms. In 1992 fully three-quarters of the total credit claimed was from this source. Of this amount, in turn, 30% was claimed by companies in the chemicals and allied products category. Pharmaceutical companies accounted for 22.1% of this amount. The remainder was accounted for by electrical equipment, transportation equipment, and machinery manufacturing firms. The breakout by the major categories of claimable expenses showed wages to be 62% of the claimed R&E expenses, supplies 20%, contract research to be 12%, and basic research payments to qualified organizations (universities and others of a similar nature) to be a mere 0.4%.

Thus, what the tax credit accomplished in the ten years preceding the OTA evaluation is that it provided assistance in defraying the in-house research expenses of a relatively narrow spectrum of corporations in the chemical industry.

What the tax credit provisions have not done, the last figure above clearly indicates, is produce a significant allocation of research moneys to universities. The inference is either that companies prefer to perform their own basic research, or that they stop doing it altogether. To the extent that the latter is the case—and it appears to more and more be the case—it must be concluded that there has been a "market failure".(*2*) In fact, it has become conventional wisdom to say, "The R&E Tax Credit represents more of a financial tool than a technology tool". (*1*) It has also been said that the R&E tax credit "is a blunt and expensive instrument that is now a tired idea politically".(*3*)

How might the R&E Tax Credit be Improved?

Using the language of economics, the problem to be addressed is the market's tendency to undersupply basic research as well as infrastructural or generic research. The means to overcoming this in the context of R&E legislation is suggested to be as follows.

• Change the definition of basic research (4) in such a way as to cover more of what industry might do collaboratively in not-for-profit laboratories.

• Give the nation's government labs equivalent status for industry funding credit purposes to that accorded universities and other not-for-profits. (5)

• Increase general credit rate from 20% to 25% and the credit for industry's joint ventures with not-for-profit laboratories from 20% to 100%. (6)

Broader Coverage. The first of these recommendations addresses a major disincentive in the existing legislation. The language of the IRS Tax Code (4) defines basic research as any original investigation to advance scientific knowledge *not having a specific commercial objective* (emphasis mine). Branscomb (3) prefers to think of basic research as including both "basic scientific research" and "basic technology research," and he maintains that in reality both are "relevant to commercial as well as public purposes." Universities are unquestionably the principal sources of basic research. However, it is well recognized within industry that government labs represent a source of both basic scientific and basic technology expertise in certain areas greater than typical industrial labs can afford to build or acquire on their own. Hence, the R&E tax credit legislation could become a vehicle for promoting industrial lab investment in both resources, not just universities.

An expanded definition of basic research would simply recognize what has been accomplished by years of Congressional effort to write and perfect legislation giving to universities, and indeed government labs, the rights to apply for, own and then license patents to interested commercial entities. Provisions under the 1980 Bayh-Dole Act as amended in 1984, the 1980 Stevenson-Wydler Technology Innovation Act as amended by the Federal Technology Transfer Act of 1986, the National Competitiveness Technology Transfer Act of 1989, as well as other acts, amendments and executive orders, have allowed universities, and more lately government labs, to in fact become technology resources for the private sector. That is, they have become licensers to private sector interests—at fees which in some cases have amounted to many millions of dollars over the life of a patent. The Association of University Technology Managers, for example, reported total patent income from member institutions of $260 Million in 1992. This included license fees, option fees, annual minimums, termination fees, and running royalties.

That there has been a massive transformation of university and government lab research culture resulting from this cannot be missed. Indeed, both types of institutions have sizable technology transfer organizations consisting of patent attorneys and other administrators engaged in setting up agreements and negotiating terms with industrial counterparts. Out of this there has arisen a technology transfer infrastructure which includes numerous regional, national, and international conferences at which "market-ready technology" is made available to corporations of the world to develop and commercialize. Yet, it is no secret that inventing commercializable technologies is not what universities and government labs are particularly good at, nor is it, in general, what companies want, either from the universities or the government labs. What industry wants, as Roessner pointed out some time ago (7) in reference to government labs, is something other than technology transfer, something rather more like idea transfer. Stated another way, the preferred relationship is one of shared interests and complementary talents leading to co-inventions in areas of mutual interest.

As a consequence, funding from industry for university or government lab research that has patents-in-prospect is more often the rule today. Moreover, when funding is provided by corporate research organizations for such collaborations, it will typically be based on an agreement which specifies that the funding source has the right of first refusal to license any resulting patent and to commercialize the invention. This kind of arrangement takes nothing away from universities or government labs. Rather, it adds to the mix a component of skills that neither of those organizations typically has in its portfolio.

Equivalency of government lab status with universities *vis-a-vis* industrial labs would accomplish another improvement for government labs. Their existing equivalency to private contract labs requires any R&E expenditure to be discounted by 35%. This is Congress' "rough proxy for the fact than many overhead costs and support staff activities would not qualify for the credit if the research were performed in-house." (1) It has been established by a study for the Government-University-Industry Research Roundtable, however, that overhead costs are essentially the same for universities and government labs.(8) At the very least, therefore, government labs should be exempted from the contract lab discount.

Higher Credit. The formula for calculating the Research Tax Credit (RTC) is recommended to read (cf. Appendix A):

$$RTC = 0.25\,QRC + 1.0\,BRC$$

This recommends a change in the 20% credit level to 25% for the Qualified Research Credit (QRC) and to 100% for the Basic Research Credit (BRC). The basis for the first recommendation is that it will take something such as this in order to make the credit both a technology tool and a financial tool. The question is whether this increase would really suffice. It is the amount of credit presently offered in the State of California, and it equals what was originally provided in the 1981 R&E legislation. The modification last year providing the Alternative Investment Credit (AIC) surely will add more companies to the ranks of claimants. However, it would not appear likely that, by itself, the AIC will raise the credit to the level of R&D planning considerations.

The basis for the second recommendation is that only with such a significant change could the credit legislation hope to become an incentive for basic science investment by corporations. As with the previous recommendation, it is uncertain whether a change of this magnitude, though large, would suffice. The magnitude of change suggested is not unprecedented. Australia provides 150% deductibility for all R&D investments. (1)

Expected Benefits

If enacted, these three recommended changes—broader definition of basic research, increased institutional options, and increased credit levels—would have the effect of testing the theory that private sector research portfolios cannot be altered by a credit mechanism. Only a family of such changes would seem to be able to address previous criticism adequately. Together, the proposed changes have the potential of making the credit not only a financial tool for industry, but also a science and technology support mechanism for the non-for-profit sectors of the chemical research enterprise.

The latest survey results published by the Industrial Research Institute (9) reveal that industrial investment in basic research has not gone out of fashion altogether, as some predicted in the early years of this decade, but in the form of university research support is expected to rise, albeit modestly. The implication is that the previous years of cut-back were an R&D investment "correction," not an irreversible change. Thus, another benefit of the proposed changes is that they would give further momentum in the direction it appears industry is prepared to go.

More to the point still, the proposed changes would promote further the natural intermingling of scientists and engineers within the public and private sector organizations. This would go beyond employee transfer, another option for increased interaction, because it would be based on the foundation of mutually interesting project opportunities. This would permit accomplishing the goal of cross-cultural acclimatization *en passant.* Moreover, making changes of the kind proposed is preferable as well to direct funding options such as the Advanced Technology Program (ATP) and preferable also to the indirect funding of projects through Cooperative Research and Development Agreements (CRADAs).

The chemical research enterprise will benefit to the extent that it evolves toward more collaboration across sectors, thereby making better use of its aggregate intellectual capital and physical assets. What is in it for researchers at universities and government labs is the synergy that may be introduced into their own pursuit of discoveries and inventions that are not only patentable but also more readily commercializable. What is in it for industry is the greater prospect that the unique and necessary assets of the other sectors not only will not be lost, but will thrive and become more affordable, hence, in essence more available to industry.

Government's role in the future of the chemical research enterprise is therefore pivotal. At a most propitious time in history, it has the unique capability and opportunity to promote survival-value cooperation among all three enterprise components in a way that none of them could hope to accomplish on their own initiative.

Acknowledgments

The author wishes to thank the following for helpful discussions: P. N. Doremus, Technology Administration, U. S. Department of Commerce; R. B. Hill, Senior Director, Taxation, Chemical Manufacturers Association; G. Tassey, Senior Economist, National Institute of Standards and Technology; and T.J. Kehoe, Department of Taxation, Procter & Gamble Co.

APPENDIX A

REGULAR RESEARCH AND EXPERIMENTATION TAX CREDIT

As a means of developing a sense of the overall benefit, one might imagine a corporation's R&D expenditures to be $800MM, of which $400MM is expended for qualified R&E activities. In order to calculate the regular tax credit using the equations below, we must first calculate the average ratio of qualified research expenses and gross receipts for the base years 1984-88. Typically, this is of the order of 1.7%. Next, if the corporation's average of gross receipts in the four years preceding the year for which the credit is being calculated is $20MMM, then the corporation's tax credit threshold for qualified research expenses is $340MM (=.017 x $20MMM). The excess of current year qualified research expenses over base period expenses is $60MM(=$400-340MM). Hence, the nominal research credit is $12MM (=.20 x $60MM). When this amount is added back to the taxable income and taxed at 35%, the net benefit to the corporation is $7.8MM (=$12MM - (.35 x $12MM) = $12 - $4.2MM = $60MM x .13). This is in the absence of any qualifying basic (university) research payments.

Assume, for example, that $5MM of the $400MM is in the form of basic (university) research payments, and the corporation's base (threshold) amount of basic research payments is $2.5MM, then the excess of $2.5MM over the base amount would yield a net benefit of $325M. (The below-base amount is also permitted to be used in the benefit calculation, although at a discounted rate (65%), so the true benefit is actually $436M in this example. This is a kind of double incentive for industry to have research performed at a university.)

$$RTC = 0.2(QRC + BRC)$$

$$\text{where} \quad QRC = QRE_t - QRE_o$$

$$BRC = BRP_t - BRP_o$$

[$_t$ denotes "taxable year" and $_o$ denotes "base period"]

$$QRE_o = \frac{aveQRE_y{'}}{aveGR_y{'}} \times aveGR_y{''}$$

[$_{y'}$ denotes the five year period 1984 through 1988]
[$_{y''}$ denotes the four years preceding the year for which the credit is being determined]

$$BRP_o = MBRA + MOEA$$

$$\text{where} \quad MBRA = \max\{.01 \times aveQRE_y{'''}, aveBRP_y{'''}\}$$

$$MOEA = CPI \times aveNDUC_y{'''} - NDUC_t$$

[$_{y'''}$ denotes the three year period 1982 through 1984]

Symbol Key			
AIC	= Alternative Incremental Credit	**MOEA**	= Maintenance-of-Effort Amount
BRC	= Basic Research Credit	**NDUC**	= Non-Designated University Contributions
BRP	= Basic Research Payments	**QRC**	= Qualified Research Credit
CPI	= Consumer Price Index	**QRE**	= Qualified Research Expenses
GR	= Gross Receipts from U.S. Tax Return	**RTC**	= Regular Tax Credit
MBRA	= Minimum Basic Research Amount		

APPENDIX B

ALTERNATIVE INCREMENTAL TAX CREDIT

Assuming the same parameters as in the example calculation in Appendix A and using the formula provided below, the results are as follows. The threshold is now $200MM (vs. $340MM), so the excess over the threshold is $200MM (vs. $60MM). The first $100MM of the excess produces a credit of $2.2MM (=.022 x $100MM). There is no further credit earned in this example because expenditures did not exceed 2% of the $20MMM average yearly gross receipts. The total benefit by this calculation is therefore $3.85MM. This compares to $12MM by the regular credit calculation.

The alternative investment credit calculation proves more favorable when a corporation's R&D expenditure either does not exceed its threshold amount as defined for the regular credit calculation, or does so by a relatively small amount. An example would be if the corporation in the example above had qualified research expenses of $350MM instead of $400MM.

$$AIC = .0165 QRE_t' + .0220 QRE_t'' + .0275 QRE_t'''$$

$$\text{where} \quad QRE_o < QRE_t' \le 1.5 QRE_o$$

$$1.5 QRE_o < QRE_t'' \le 2.0 QRE_o$$

$$2.0 QRE_o < QRE_t'''$$

$$\text{with} \quad QRE_o = .01 \times GR_y''$$

$$\text{and} \quad QRE_t = QRE_o + QRE_t' + QRE_t'' + QRE_t'''$$

[$_t$ denotes "taxable year" and $_o$ denotes "base period"]
[$_y''$ denotes the four years preceding the year for which the credit is being determined]
[$_t'$, $_t''$ and $_t'''$ denote the first, second, and third credit tiers in taxable year $_t$]

Symbol Key			
AIC	= Alternative Incremental Credit	**MOEA**	= Maintenance-of-Effort Amount
BRC	= Basic Research Credit	**NDUC**	= Non-Designated University Contributions
BRP	= Basic Research Payments	**QRC**	= Qualified Research Credit
CPI	= Consumer Price Index	**QRE**	= Qualified Research Expenses
GR	= Gross Receipts from U.S. Tax Return	**RTC**	= Regular Tax Credit
MBRA	= Minimum Basic Research Amount		

Literature Cited

1. Office of Technology Assessment, Congress of the United States, "The Effectiveness of Research and Experimentation Tax Credits," OTA-BP-ITC-174, Sept. 1995.
2. Tassey, G., *Issues in Science and Technology,* Fall 1995, p 31.
3. Branscomb, L., *Issues in Science and Technology*, Spring 1997, p 41.
4. Code Sec. 41(e)(7)(A).
5. Code Sec. 41((e)(6).
6. Code Sec. 41(a)(2).
7. Roessner, J. D., *Issues in Science and Technology,* Fall 1993, p 37.
8. Government University Industry Research Roundtable, "The Costs of Research: Examining Patterns of Expenditures Across Research Sectors," Mar. 1996 .
9. Wood, R., "Industrial Research Institute's R&D Trends Forecast for 1998," Research Technology Management, January-February 1998, p 16.

Government and Media:
Support for the Research Endeavor

12

Setting Sustainable Research Priorities

Hon. Robert S. Walker

President, The Wexler Group
1317 F Street, N.W., Washington, DC 20004

The need for the science community to understand the federal budgetary process in order to assure sustainable research priorities is discussed. Government support for research has become more mission oriented and often does not differentiate between technology and basic scientific research. Policymakers need to understand science policy. Often scientists seek funding through appropriations but these can change each year with new policy decisions. Budget authorizations, however, assure scientists long term sustainability of their research. Applying the tax policy in innovative ways can increase the national investment in science, for example, by making the R & D tax credit permanent and also applicable to industry-university partnerships. "Big science" will have to be pursued through international consortia and in the context of the whole economy by bringing resources from a variety of areas.

Public Policy Different from Science

Good afternoon. I am Bob Walker and I am delighted to be with you this afternoon. I am going to talk to you about public policy. I do not pretend to know science. I got drafted for the Science Committee when I first went to Congress. They called me up one day when they were picking committees and I was the new freshman congressman. They said, "You are going to be on the Science Committee." I said, "Fine". I went to the congressional directory to find it, to try to figure out what this meant. When I actually got on the committee though, I figured out that this was an exciting era. We are really defining the future and I got a chance to glimpse over the horizon. So it was in Congress that I kind of majored in science and I imagine that my college professors and high school teachers were flipping over, thinking that I had anything at all to do with science policy. But it was really a fascinating learning experience and some of the people here today were amongst those who helped educate me along the way.

What I want to try to talk to you about today is the fact that public policy as it relates to science sometimes defies what you think should be the case. For example, the other day, as I was pumping gas, the guy beside me wearing a sweatshirt with the slogan, "When all else fails, manipulate the data."

Well, I know that in the science world, we are talking disaster if you do that. In the public policy arena, that is exactly what you do. When you cannot get the answer you want, you simply manipulate what it is you need and go on. Thus, it sometimes makes it difficult for you scientists to figure out just what it is that is going on as we try to set sustainable research priorities in the public policy arena. So, based upon what is really happening in the public policy arena, I want to do is talk to you about where I think this business of setting priorities in science is going to go in the near future and, hopefully, maybe down the road, as well.

Science in the National Interest

First of all, there is going to have to be a real sorting out of this whole process of what we mean by science in the national interest. It is not just a question now, of whether or not you are a part of the federal budget, but what else is going on in the totality of society, in particular in the totality of the investment community, that is spending in science. How do you blend science and investment? That is not just a federal budget question. One of the things we are going to have to work out is what is federal and what is not.

Now, I would submit to you that it becomes pretty clear that you need to have basic research conducted as a part of the Federal government's program. It is going to be real hard to find investment money that will be sustained over long periods of time for basic research. If you are going to set a priority, the fundamental priority that you have to have in terms of government policy is that government is going to do basic research.

Now, I realize that there is no clear delineating line between basic and applied in a lot of these things. They blend with each other and it is very difficult to figure out where one ends and the other begins. But, it is clear that fundamental research, with its long duration and learning curve, will probably not get done, if it is not done at the federal level. So, you are going to have to budget for that at the federal level.

Distinguishing between Science Policy and Technology Policy

Now you have to begin to distinguish between science policy and technology policy. That is a distinguishing characteristic, which is not necessarily clear in government. There is a tendency to lump together science and technology into one pot and say all of this money is going toward research. The problem for you scientists is that if support begins to slide toward technology, at the expense of basic research, there will be underlying and fundamental questions for the sustaining of the basic research priority.

Let me give you an example. In my view, the more you slip toward technology, the more political the decisions become. If you read the newspaper this morning, there were a couple of line item vetoes by the President yesterday. Mainly it was the science and technology oriented items that were lined out, because the administration said, "What we are doing is getting rid of corporate welfare." But, when some of us in Congress argued about the ATP program, that some of that was corporate welfare, the administration reacted and said, "Oh, no, that is not corporate welfare because those are our priorities."

The point is that the vetoes yesterday seemed to say that if it is in our budget, it is okay, — it is good technology, it is good science, it is good research at that point. However, if it is not in our budget, it is not okay, it is not good research, and it is not good technology. I suggest to you that at that point, that is a political decision. It is not a peer review decision, nor a science decision, but a purely political decision. And when you start to get political decisions governing the priority of science, you have a problem. You have a lot of people who make those decisions based upon no scientific criteria, whatsoever.

Politics of Technology Policy

The other point I would add is that the more policy slips towards technology, the more you are in the process of picking winners and losers within the overall economy. That's a real problem, because right now there is a lot of investment taking place in very high tech and science type programs with much of it is going into relatively small firms, start-up firms.

Some of the investment is being made by relatively big firms that understand that being on the technological cutting edge is what the future is all about. But, when the government decides what it is we are going to support over what we are not going to support in technology, it is really in the process of selecting which company is going to succeed. And that is particularly true with small firms.

If you do not believe that, take a look at a program like the ATP program which was designed with the good intent of being pre-competitive technology. But, when the GAO polled the firms to determine what they thought they were getting out of this program, sixty percent of them responded that they thought that at the end of the period of the government's investment, they would have a new product they could sell.

Well, that is not pre-competitive. That is very competitive. It means these firms received government money to do something that their other competitors had to get without government money. It's a big problem.

I will also tell you that the government is always going to be involved in mission oriented research. An example of mission oriented research, is research carried out by the military because technologies and science it requires. NASA does science because of things that NASA, in its unique role, has to do. You are going to have some programs put together, simply because of government programs and government policy, such as antitrust policies.

There is a need, at times, to bring together all of the companies that may be involved in a particular sector of the economy to see whether or not they can't come up with some key solution that is needed, and to do so in a way that does not violate the antitrust laws. When the government puts such an effort together, I would tell you, it is mission oriented. It also means committing the government laboratories to the effort. While we have to figure out a way to have some interaction between the government labs and the totality of the economy, the fact is that the government labs perform their own function, which is rather unique and is important to the long-term future.

Change in Federal Decisions

The other thing that you need to realize is that there is a brand new dynamic in the way federal decisions are made. Federal decision making has changed fundamentally in the last few years. Beginning in the 1970's, and particularly in the 1980's, as a result of budgeting, we began to set priorities. The moment the budget is written, the real priorities get set. It becomes a general exercise, within which the amount of monies that are going to be spent are put in categories. The allocations are then determined by the various subcommittees of appropriation. That allocation means all the difference in the world as to how much money can actually be put in the programs in any given year, and all of that budgeting starts when those overall goals are being determined. That is very different from the way it used to be when people came in and pleaded their case to the appropriators and the appropriators figured out how much they had within the overall pot and how much money they could actually spend for something. Now there is a set of priorities which is set early. Thus, if science is not involved in the budgetary level, science is not going to be taken into account.

I have to tell you, one of the reasons I got appointed Vice-Chairman of the Budget Committee, when we took control of Congress a couple of years ago, was because Speaker Gingrich was very interested in science and he said, "I want the Chairman of the Science Committee as the Vice-Chairman of the Budget Committee." So we made certain that that priority was implemented.

You need to make certain you have other legislators who have similar concerns about science also in the budget process because we did a number of things in the course of just deliberating within the Budget Committee. I will give you one example. We were sitting at the Budget Committee one day and someone came up with the idea that we should get rid of the overhead cost of the universities. We would save a billion dollars by taking out the overhead cost that goes into science. I said, "Wait a minute. That just comes out of the hide of science. I do not care how you define it. It comes out of the hide of science. You can just define it as overhead, but the fact is, it is just going to come out of the overall amount of money that goes to university related science." We talked it through for a little while and eventually I got that taken off the blackboard, so that it was no longer one of the items that could be considered. But unless you have someone in the budget process that understood the issue, it could sound to the others like an easy thing to do, for they argued that we are not going to touch science at all, we are just going to take it from the overhead costs. That is why somebody who understands science policy needs to be in a position to help make those decisions.

Why Authorize

Furthermore, there is a real need for authorizations. Now I am getting into the real technical aspect of what Congress does. There is a difference between authorizing and appropriating. Authorizing means that you pass a bill that sets the policy framework for the spending ultimately to take place. Appropriating means that you actually put the dollars into a program effort. In recent years, there has been less and less authorizing and more and more appropriating, and as a result, a lot of people in the science communities have decided that where the action is in appropriations and so therefore we will go in and get the appropriation, and we won't worry much about whether or not the program is authorized.

But the problem is, that if you do not have authorization, you have no long-term sustainability because it is the setting of the policy that then assures that you have some ability to stay in it over a long period of time. No appropriation is for more than one year.

The Appropriations Committee operates on one-year cycles and that allows them to take control of things and manipulate them year by year. In the United States Senate, many Senators serve on both the Authorizing Committee and the Appropriating Committee and they prefer to do appropriations rather than authorizations because it allows them to come back and revisit their decisions every year. That is a disaster in basic science.

Therefore, you need to begin to encourage a process that assures that you get a policy framework for the appropriating. That means each year we have to pass NSF authorization bills, NASA authorization bills, and the Department of Energy authorization bills. These must be done to get the right kind of priorities. It is also the reality now. As I mentioned just a couple of minutes ago, you now have this new item called "line item veto". That affects only appropriations.

Line Item Vetoes

But it does mean that the Administration now has an ability to second-guess the Congress on every line item of spending in the budget. And, if you look at what happened the other week, when the defense appropriation bill was line item vetoed, there were a whole series of line item vetoes, most of which were in space or in research. Maybe some of them were bad programs. When I looked at them, over half the money was in R & D related efforts. The judgment in the line item veto is that science takes a hit. You have to be worried about that. As long as you have a philosophy that says that the decision is going to be made as a result of an appropriation rather than an authorization, you have very little way to go back and say no to a line item veto because this is a very good project. Only if you have an authorization in place, a national policy in place, can you say this is in line with what we decided to do as part of national policy.

Better Science Through Better Tax Policy

Therefore, if you are going to have sustainable priorities in the future, you ought to begin to think of innovative ways to use tax policy as a way of increasing the national investment in science. Now this is outside the federal budget criteria. It is more toward trying to make certain that more money moves into the science area. For example, what if the research and development tax credit, first of all, was made permanent so that everyone could count on it. It would make a big difference in what businesses in the boardroom were deciding about research investment of their own.

But, what if we also took the R & D tax credit and made it applicable so that if a company wanted to invest in basic research at a university, it could do so and write it off against its R & D tax credit? All of a sudden this would encourage partnerships between academia and business that could have a positive impact on the business community. These partnerships might also move into other things. But fundamentally this would probably increase the amount of investment that would go into science overall. That would not be a budget related decision, it would simply be a policy decision, and the business community would decide how the R & D tax credit would be used.

Impact of the Global Economy

Finally, in thinking in terms of sustainable priorities, we must consider the overall nature of the global economy. The global economy means that everything done here has an impact elsewhere, everything done elsewhere has an impact here. It is the reality of the information revolution and it is the reality of the way science works. Science for the good of all is a very, very promising piece of what the next century holds. But it also means that we need to think about our investments in terms of that global economy. For example, it seems to me that if we are going to do big science projects in the future, the kinds of things such as huge fusion reactors or other kinds of very big science, it is going to have to be done internationally.

Science will have international consortiums, international investment, with many countries participating, if it is going to do those big projects. So, anybody who is suggesting big science projects that do not have an international context, is probably going to end up a loser in the public policy arena. The space station is somewhat of an example of what I am talking about here as a truly international project.

You have a consortium of countries that have come together and decided to fund a space station, many for their own reasons. It is the kind of thing that happens. The LDC in Cern is another example of many countries joining in investing. It's big science being done in an international context. Now, that will raise a lot of political questions. It means that countries will have to sort out the reasons for their investment, but never the less, it seems to me, that it's the only way you are going to get those investment dollars in the future.

My point is this: you as scientists are going to have to view science in the context of the whole economy. You are going to have to see it in national interest, but in the total national interest, where you try to bring in resources from a whole variety of areas. And you need to look at the Federal Government as the place where you get the investment in basic research. If, in fact, you focus on those goals, then we can sustain the priorities for science.

13

Chemistry: A Look Ahead at the Media Coverage

Robert F. Service

Science, **American Association for the Advancement of Science**
1200 New York Ave., N.W., Washington, DC 20005

Media coverage of general science topics such as chemistry has declined in recent years. This report looks at some of the reasons for this decline, how the Internet is changing the landscape of the media's coverage of science, and what the chemistry community can do to bolster coverage of the field.

Introduction

It's a pleasure to participate in this symposium with such distinguished company. We've heard a lot today about some very exciting fields of research, the directions where chemistry is likely to proceed in the future. I'd like to spend a few minutes sharing my views on where the media coverage of chemistry is likely to go in the next several years.

Changing Coverage

First a bit of history. Media coverage of chemistry and science in general has been changing in recent years. There is little information breaking down how much media coverage there is of different fields of science, such as chemistry and physics. But one piece of information that is most likely reflective of trends in this coverage is the number of science sections in newspapers. As recently as 1989, there were over 100 newspapers across the United States that carried separate news sections devoted to covering science. By 1992, the number of such sections had dropped to around 50, and by 1995 that number had shrunk to about 35. What is more, the focus of these sections has changed considerably. Of the science sections that remain, many have shifted the bulk of their coverage from stories about a wide variety of scientific disciplines to stories focused largely on "health" and "medicine" (*1*).

Media Coverage of Science

♦ **Overall science coverage is down**
- 1989: Over 100 newspaper science sections
- 1992: ~50
- 1995: ~35

♦ **Coverage has changed from general science to "health" and medicine"**

Why the shift in coverage? There are a number of reasons. Most importantly, few science sections turn a profit from advertising. On top of this, newspapers and magazines have been hit with a considerable rise in paper costs in recent years. Together, these have made it difficult for publishers to justify the expense of news sections that don't pull their own weight. On the television side, in recent years network news programs have suffered a steady decline in viewers. In an effort to halt this decline, news directors run stories that they believe will appeal to the widest audience. More often than not, that tends to be stories about crime and celebrities than science.

So, the big picture is a decline in mainstream media's coverage of a wide variety of general science topics. One result of this trend is that stories about chemistry—in newspapers anyway—must now compete head to head for space in the main sections. Space is tight and it's impossible to run all possible stories. It's a bit like grant funding in that way: There's always more good ideas than money. To get covered, chemistry stories must now beat out not only stories about physics, health and medicine, but national and international news stories as well.

Deciding What to Cover

So how do reporters and editors go about deciding what to cover as news? There are a number of different types of news stories; daily events that take place, trends, and so on. This is true for stories about science, just as it is for those about crime. What gets a story covered is a combination of its importance, interest and timeliness. Reporters justify each story to both their editors and readers. Again, the situation here is not much different from writing research grants. Researchers must give funding agencies a reason to support their proposals, just as reporters must clearly explain to readers why they should read this story now.

In my mind there are a handful of different types of science stories that tend to get covered most frequently. This list is by no means exhaustive. One type of story is about fundamental scientific developments that take place. Such stories are by nature few and far between, since true fundamental breakthroughs are actually quite rare. In this category I would include stories such as the discovery of buckyballs and the top quark in physics. These were stories that unexpected or not had profound influence on their respective fields as a whole. And it's that sense that something new has been learned that's likely to affect an entire discipline if not the lives of readers that gets the story covered.

A second common type of story broadens readers' understanding of how they fit in their world. In this category I place a lot of stories about disciplines such as archaeology and astronomy. I'm often asked why these topics get so much media coverage, while far larger disciplines such as chemistry receive comparatively little. I believe that it's because these types of stories help give readers a brief look at where our species came from and what our Universe looks like. Most people are naturally curious about such topics.

A third type of story—and one that pertains to a bulk of coverage about chemistry—has to do with events that take place that are likely to have an impact on the lives of many readers or viewers. In this basket, I place stories about the development of new medications, health research, and the latest technological wonder.

Additional categories include stories about controversies in a scientific discipline, battles over funding, developments in science and technology companies, and profiles of companies and researchers. One topic that is notably absent from this list is stories about techniques. Researchers spend their lives in the lab and its typically a breakthrough in technique that leads to important results. But as important as this is to researchers, it's not nearly as important to readers. They want the results first and information about how those results came about second. What is new and what is the impact of that? That's the news.

Deciding What to Cover

♦ **Reporters must justify each story**

♦ **Common stories**
   - Fundamental breakthroughs—buckyballs
   - Fitting into the world—archaeology, astronomy
   - Life impacts—new medicines, products
   - Controversies, funding fights, personalities
   - Company news, profiles

♦ **Stories on techniques rarely sell**

Here Comes the Web

So, coverage of science is down. Yet, there are still many types of science stories that get covered. But there is one other factor in the news business that is beginning to change this equation that I have yet to discuss—the web. The Internet has begun to drastically transform the news business. By now, most news outlets have created new versions of their publications and news programs in cyberspace. Many of these web sites contain stories that never appear in print or on television. The good news is that this may therefore be increasing the number of science stories that news organizations are able to cover. *Science*, for example is a weekly magazine. But about a year ago we started a online news service called *ScienceNOW* that covers new developments in research and science policy every day. In just the first year, we've run several hundred stories in *ScienceNOW* that never appeared in print.

Yet, the web may have a downside as well. One analyst of the news business that I spoke with, told me that the web may wind up removing even more non health and medicine coverage from network television news (2). Historically, networks have tried to balance the types of stories they do over the long term. They strive to report on a variety of disciplines in their science coverage. But with ratings pressure, the current drive is to just cover topics that have the most viewer interest, such as stories about health. Now, with the web, news outlets will be still be able to cover a wide breadth of fields, by putting stories perceived to have less viewer interest on the web, and broadcasting only stories with a broad appeal. Broadcasts coverage of general science in this case would fall. It's understandable in a free market world. But is a story about the health effects of vitamin C more important than the development of a new fuel cell technology, that could provide clean energy to millions?

The Web Factor

♦ **Most media outlets have web sites**

♦ **Web allows coverage of additional stories**

♦ **But news outlets may push all but "top" science stories to the web**
  - No need for diversity of coverage
  - Niche interest stories to the web

♦ **May reduce general science coverage**

Improving Matters

Can this situation be improved? My answer is an emphatic yes. First of all, I don't think the news coverage of chemistry is absent. Actually, I think chemistry gets covered quite a bit. Reporters just never call it chemistry. Here, I'm thinking about stories on products that are important in peoples' lives, such as new cars, fashion, and medicines. Developments in chemistry are vital for all these fields. Little news coverage focuses on the research. Rather it looks at the products that are the result, since it's the products themselves that make a difference to most people, not how they found their way to store shelves.

But beyond this point, I think there are things that the chemistry community can do to help news organizations improve their coverage of research in the field. Few reporters are trained in chemistry, or even science for that matter. Without that training it's not easy to know which developments in the field are important and therefore deserve coverage. That's where the community comes in. I believe the chemistry community needs to take a more proactive role in helping reporters determine which developments are significant.

There are a few ways to do this. The simplest is the everyday press release. These are often distributed to reporters in advance of the publication of a paper or a report given at a meeting. This advanced notice gives reporters a heads up, highlights the impact of why this story is important, and gives the reporter time to report and write a story so that it can coincide with the publication of the research. This standard approach often works. But it can be quite spotty, because often few of the public affairs people who write the releases are trained as chemists either.

A twist on the conventional press release is an approach taken by the American Institute of Physics (AIP). They have trained physicists that peruse upcoming journals and send out weekly alerts to reporters of what they consider to be the most important pieces of research. Included in each item is enough background to bring the reporter up to speed and understand the item's significance, as well as contact numbers for the scientists and outside experts in the field. This works, I believe, because AIP has made a commitment to highlighting cutting edge research week in and week out. That reliability results in more physics stories being covered more often.

For a final way to improve chemistry coverage, I think the chemistry community can follow the lead of another community of scientists, molecular biologists. Each summer, the Woods Hole Institute's Marine Biology Laboratory invites a handful of journalists to spend a week or more at the Institute working side by side with top researchers in the field. Over the week, journalists take courses, go out in the field and work in the lab. They learn about cutting edge research topics first hand. This helps them to better recognize important developments, provides them with contacts that they can talk about future stories, and gives them an appreciation for how science is done. This isn't an approach that pays off immediately. But over the years, as more and more journalists go through the workshop, the quality of coverage of marine biology, molecular biology and neuroscience continues to improve.

Making a difference

♦ **Chemistry is covered: Cars, fashion, drugs**

♦ **Reporters must be sold on impact**

♦ **One line in a journal article is not enough**

♦ **What works?**
- Woods Hole marine biology workshop model
- Physics News Update: weekly update of upcoming journal articles highlighting impact

To summarize, I think this is a very dynamic time in the media coverage of science. The coverage of general science appears to have fallen since the 1980s. But with the advent of the web, there are new outlets for science reporting. This may allow for increased coverage of topics such as chemistry research that don't get as much attention in the papers and broadcast news. Finally, the participation of the chemistry community can help reporters and editors gain new insights into the discipline and encourage them to take a deeper look at progress in the field.

Literature Cited

1. Haycock, D.A., "Who Killed the Science Section?," August 8, **1997**, URL http://biomednet/hmsbeagle/1997/14/people.pressbox.htm.members
2. Robert Logan, personal communication.

Chemistry: The Central, Creative, and Enabling Science

Comments and Discussion

14

Challenges for the Future of Chemistry: A Panel Discussion

Ronald Breslow

Department of Chemistry, Columbia University
566 Chandler Laboratory, New York, NY 10027-6948

The contributions of chemistry to human welfare are discussed, along with a description of some challenges for the future of chemistry. Topics covered include the magnitude of the chemical enterprise, and past contributions and future prospects in the areas of health, life, materials, the environment, computers. electronics, catalysis, synthesis, structure determination, and reaction mechanisms.

Chemistry, the Central, Useful and Creative Science

I want to talk a little bit about the challenges for the future, not in terms of politics—which is too knotty and difficult a subject for us to deal with—but in terms of science. Let me tell you a bit about what I, and a number of other very thoughtful people, think are some of the interesting challenges for chemistry.

I published a book with the American Chemical Society last year called *Chemistry Today and Tomorrow—the Central, Useful, and Creative Science. (1)* I want to talk about the "tomorrow" of chemistry, but first I do want to say why I called chemistry the central, useful, and creative science. You can find various centers in the continuum of scholarship, but certainly, chemistry abuts on so many fields that it is at least one of the central sciences.

The creative aspect is really quite special to chemistry. There is no such field as synthetic astronomy, for instance, or synthetic geology. People aren't making new universes or new worlds to see how they can compare with the one that we happen to have. We chemists extend the natural world all the time, with tremendous effect.

In 1965, one million new compounds were registered in Chemical Abstracts, compounds chemists created that had never existed before. Well over ninety percent of all known chemicals are substances that chemists themselves have created over the years. So, chemistry certainly is a creative science, and that's one of its charms. I will talk a bit later about chemistry as the useful science.

Health, Life, Materials, and the Environment

Chemistry is essential to the health and life of our nation. Medicinal chemists are the people who invent and create modern medicines, although the public sometimes believes that physicians do that. The importance of modern medicines is hard to overstate. Consider the fact that life expectancy at birth has increased from 47 years in 1900 to 76 years today, mostly because of medicines invented by chemists and sanitation procedures also invented by chemists. The chemistry of life is also a very important and exciting area. Most thoughtful biologists believe that we are going to understand living processes by explaining them in chemical terms, at least to a large extent. (2)

Chemistry is the useful science in its contribution of materials of various kinds, that play a role in housing, transportation, national defense, electronic devices, etc. (3)

Chemistry is also finding solutions for environmental problems. A lot of people would say, "Yes, chemistry is the useful science, but ..." What comes after "but" is, "you are wrecking the planet". We can't afford that. There are some very exciting things going on in that area. Some of them were mentioned already this morning by Paul Anderson. The Responsible Care program of the Chemical Manufacturer's Association is very exciting and, frankly, very under reported. Most of the public has no idea of the tremendous change that has gone on in an industry that is fundamentally pledged to make chemicals in such a way that the manufacturing process will not cause planetary damage, and that the products themselves will not be a problem.

One of the interesting problems with chemistry in the environment has to do with persistence. We all knew, at one point, that it was very desirable to make compounds as stable as possible. You'd make a refrigerant that could sit on the shelf for a thousand years. Or, if you wanted to get rid of the insects in some planted area somewhere, you'd put down an insecticide that would last forever so that you'd only have to put it down every hundred years or whatever. We now know that the challenge is to make things with the correct amount of persistence; they have to last only long enough to do their jobs, then be degraded in a harmless way. It's a challenge that chemists are addressing now, and will, in fact, successfully solve. In some areas, they've already done it. (4)

Computers, Synthesis, Structure and Mechanism

Some people know about the tremendous contributions that chemistry makes to electronic materials, but the role of computers in chemistry is also very exciting. Many young people are very interested in computers. They usually don't realize that one of the best ways to make a living, doing computation, is to be a computational chemist. Every drug company I know of has a computational group designing drugs. (5)

The way chemists create a new molecule is the way architects or construction people create new buildings. As we know, you don't throw a whole bunch of bricks and window frames together and hope that somehow they will come together in the right shape. Nobody thinks that's the way to make a building, and we know that's not the way to make a chemical. But, the public doesn't really understand how we do it. (6)

What are the ideas behind the strategies and tactics involved in synthesis? It's closely related to all kinds of tactical planning, like how to plot a decent chess game. You have to think about the end game, the middle game, the beginning, and how to put it all together.

The Chemical Process Industries

What then is chemistry? What do chemists do? What was some of the earliest chemistry? Of course, it goes back to cave man times. If you define chemistry as, among other things, performing changes, where you turn one chemical into another, people were turning wood into charcoal, turning fat into soap, before recorded history. (7)

Then, of course, the alchemists did this wonderful thing that we call "shake and bake chemistry". We all laugh about it. You mix things up, heat them, and see what happens. Think of what came out of all that. Who could have figured out that they could make glass by mixing the components and heating them up? That style of chemistry went on for a long time. There's still some of it going on, but, of course, we really hope that we are doing it in a much more classy way now.

How large a field is chemistry? There is just one comparison I want to mention. Paul Anderson, this morning, mentioned the chemical industry, which is about ten percent of U.S. manufacturing. But that's only the self-defined chemical industry and does not include, for instance, the pharmaceutical industry. It does not include pulp and paper, or petrochemicals, or the rubber industry. If you put all those together you have a group that can be defined as the chemical process industries.

The chemical process industries are those that carry out a chemical change of some kind in the course of making their products. Of course, the pharmaceutical industry does that, pulp and paper does that. As you know, you have to do chemistry to get the lignin out of wood. In the chemical process industries the processes are designed by chemists, performed by chemists, and monitored by chemists.

The chemical process industries are over thirty percent of U.S. manufacturing in terms of value added, which seems to be the only sensible measure. In other words, what was the value of what you started with? What is the value of what you ended with? Over thirty percent. Automotive is about seven percent. Mining and utilities are down in the single digits as well, six or eight percent. The only other industry that's in double digits, besides chemical process, is a category called "all other". And, that's only ten percent.

So I think we can make a case for chemistry being the useful science— its contributions to our economy, and our standard of living, are really quite fantastic.

Challenges for the Future: Medicinal Chemistry

I want to talk about what is left for chemists to do. I call this "Some Challenges for Chemistry". I asked a very thoughtful group of chemists, including some of the people on the panel here, to make suggestions about what they thought was still left to be done, and I incorporated that into the book.

Medicinal chemistry is a very important area. Think of what we don't know how to do yet. We don't have effective anti-viral medicines. Most pharmaceutical companies have people trying to invent them. We worry every year about an influenza epidemic. Where is the compound that can cure the flu, as distinguished from the vaccination that may or may not fend it off? We don't have those medicines yet. HIV, of course, is viral, but there are many of these viral diseases. We don't know how to treat them.

Resistant organisms, what a problem there. We thought we had decent antibiotics and now suddenly dangerous bacteria are immune and are popping up again. Cancer, stroke—stroke is a problem that people have been addressing, to find a medicine so that after, let's say, a mild accident of some kind or a head injury, you can stop the deterioration of nerve tissue that often follows it. Other people are trying to make compounds to promote the regeneration of nerve tissue when it has been lost or damaged. Those are both very important problems that a lot of drug companies are worrying about.

Heart disease, Alzheimers, osteoporosis, obesity, genetic defects, schizophrenia, diabetes, arthritis, these are all diseases that drug companies are working on. There is still a lot left to be done. If possible, you'd like to have a small orally active medicinal molecule you can swallow, and that means normal medicinal chemistry is going to continue to play a major role.

Drug delivery is an important problem, as is the challenge of developing bio-compatible replacement materials. Many times we replace bone by taking a piece of bone out of somebody's leg, for instance, and putting it somewhere else. Can't we, by now, synthesize artificial bone with a surface that is recognizable as bio-compatible so we don't have to tranplant other bone?

Improved diagnostic methods are important. Somebody comes in with a disease to a hospital and the question is, "What disease does he have? What's the infection?" It takes several days before a bacterial culture comes up to a level where you can identify it, but bacteria have cell surface antigens and specific DNA that should be instantly recognizable. Chemists are working on that.

Challenges for the Future: Food, Materials, the Environment

Food chemicals have to be safe and effective; there is still work to be done in this field. Improved structural materials have great potential. Think of what we do with a bridge. We make the George Washington Bridge out of steel so that it is strong enough to hold the weight of the steel that we make it from, along with the traffic. It seems a little problematic. Can't we find light, strong materials? We know we can. People are beginning to make bridges out of composites, the kind of material you find in graphite tennis rackets.   You can get the strength without the weight. We've done that in aircraft for years and people are starting to do that in other kinds of structure.

Think of what the world would be if we could have a really good, high energy density superconductor so we could transmit energy back and forth across the continent, and could generate it wherever it's convenient without resistive loss. It's a problem occupying the efforts of many chemists.

I talked about the George Washington Bridge a minute ago. It's continuously being painted because the paint doesn't last. Aren't chemists smart enough to invent paint that lasts for a hundred years so you wouldn't have to keep painting it all the time? Such paint doesn't exist yet, but it will.

In the environmental area, manufacturing has to be better. Also, it is important to figure out how to get chemical products such as refrigerants and pesticides that last the right length of time but not longer. We also need selective insecticides and agricultural chemicals that do their job, but that don't kill beneficial insects or other living things.

Let's consider alternative worlds in which we are not burning hydrocarbons in a car, but are using storage batteries that we recharge by driving into the service station. In ten minutes we recharge the battery and it's light enough to hold the amount of power that we need to drive for 300 miles. It replaces the way we currently operate. We don't know how to do it yet, but it's not for lack of work going on right now. We know that an aluminum oxygen battery, for instance, would be able to have the right weight, the right cost, deliver the right amount of energy to meet the requirements I mentioned, but nobody has a good reversible aluminum oxygen battery yet. It's a technical chemical problem that will be solved, and that would make a difference.

Hydrogen storage is an interesting problem, related to fuel cells. How can we move energy around? One of the choices is to run a hydrogen air fuel cell. George Olah, this morning, talked about hydrogen fuel cells and how they work. The biggest problem is that we don't think we can put hydrogen gas under pressure in a car. It sounds too dangerous. So, what can we do?

One plan is to have the hydrogen absorbed into something like porous nickel. Another very interesting idea is to put the hydrogen in as ammonia, a relatively safe material. Then you do a catalytic decomposition of ammonia to hydrogen and nitrogen. You send the nitrogen back to the air, you take the hydrogen and run it through the fuel cell, generating water and power. Wherever your energy was first being generated, you did the electrolysis of water and the conversion of the hydrogen to ammonia. So that would be another possible way to effectively transport electric power. There are things one can do here but they haven't been done yet. They will change the way we operate.

Nuclear waste is an interesting problem. We are going to have to end up with nuclear energy. There's not enough hydroelectric power, and we can't keep burning forever the hydrocarbon resources that we have. At some point, they will give out and we will have to go to nuclear energy. The big problem is the radioactive waste, but a lot of nuclear waste is harmless. There's a certain amount of harmful, radioactive stuff buried in a lot of dross. Chemists are trying to figure out how to isolate the radioactive part. That's a chemical problem, a separations problem. How do we get radioactive materials away from the inert material so we really don't have to bury so much or deal with so much?

These are problems still to be solved. Think of how the world would change if we finally had a hydrogen economy where we were able to move electricity around in the form of hydrogen fuel cells, if we were generating that hydrogen with nuclear plants that were safely operated, and if we could take the hydrocarbons we have and use them as chemical raw materials, which, as George Olah pointed out, is certainly what we need to do.

Challenges for the Future: Pure Science

There is a lot of pure science to be done as well. I have emphasized the practical applications so far, but there is basic research that makes all this possible.

To understand biology we must understand the chemistry of life. A relatively new challenge is understanding the chemistry of the brain. There's very good evidence that at least some memory is formed by the synthesis of chemical molecules. What are those molecules? How does that work?   When we understand it, I think we'll understand an awful lot—including, possibly, what pills to swallow to make us smarter. That wouldn't be bad. You might be able think of people for whom those pills would be useful.

We need to understand enzymes and make artificial enzymes, the area I work in. It is desirable to learn how to do what Nature does with such skill. For example, carboxypeptidase A hydrolyzes a peptide bond in ten milliseconds; we don't know how to do that, but we have to learn how. We have to learn how to run reactions in water, which is an environmentally friendly solvent, and to run reactions with high selectivity, as enzymes do.

Can we calculate protein folding? People are talking about sequencing the human genome. When they sequence the human genome, they will know the sequence of amino acids that are found in proteins, some of which will not have been identified before. What will those proteins be? The sequence is one question, but three dimensional structure is what really matters, and nobody knows yet how to take a linear sequence of amino acids and calculate how it will fold. It's an interesting and important challenge in computational chemistry.

The computer prediction of properties is an even more general question. Instead of making compounds to see which one works, we will at some point be able to compute what compound we need to make for the properties we want, and what the most efficient synthetic method is to make that compound. Work on this is going on, but it's not done yet. There is plenty to do.

There is also plenty to do in the development of synthetic methods. For example, taxol has been made in the laboratory, but by processes that are totally impractible for production. We don't yet have synthetic methodology at the point where we can really make every important molecule efficiently.

We don't really understand reaction mechanisms in full detail. I think, myself, the way this is going to go is that we will eventually compute these mechanisms. We'd like to make a movie of a reaction to show exactly what happens. We can do that now with a computer. The problem is, we don't know whether it's right. The computational methods have to get better, and they have to be checked. At some point we will be able to say " We do indeed believe that kind of calculation."

Determining the structures of big molecules is still a challenge. And constructing organized chemical systems is an interesting challenge. In the past, chemistry was reductionist in character. We would take a tree apart to see what the components are: chlorophyll, cellulose, lignins, etc. Now we are going in the other direction, saying, "What is the chemistry when we organize materials of that kind back into structures, the kind of thing one sees in a living cell?" It's a whole new area for chemistry to go into. We start thinking about not the properties of pure substances, but the properties of interacting substances. It has very exciting potential.

The origin of life is very interesting. Most scientists believe this is a chemical problem. Interplanetary and interstellar chemistry, that was referred to briefly by Dudley Herschbach, is a challenging area as well.

Message to Young Chemists

These are some of the interesting areas in which chemists have a lot to contribute, but there is one more prediction one can make with absolute certainty. All the various ideas that have been proposed in the book, all the various ideas that come from this panel about things to be done, miss some really great things that will be achieved by chemists with a different vision. And when it happens, we will all say, "What a great idea! Why didn't we think of that?"

What I have presented are not formal lists of what to do. However, they certainly indicate that chemistry isn't over yet, that there's still plenty to be done. We need to convince the public that they should care about this, and convince the media to write about chemistry occasionally. We must convince the political figures that chemistry is, indeed, a central useful and creative science and deserves the kind of support needed so it can continue to contribute to the American economy. If we can, then some of the younger people in the audience will be able to take up these problems and run with them to help move humankind ahead. Solving these problems will make a big difference in how we all live.

Literature Cited

1. Breslow, R. Chemistry Today and Tomorrow -- The Central, Useful, and Creative Science; American Chemical Society: Washington D.C. and Jones and Bartlett Publishers: Sudbury, MA, 1997.
2. *Ibid.*, pp 17-32.
3. *Ibid.*, pp 33-50.
4. *Ibid.*, pp 51-65.
5. *Ibid.*, pp 67-76.
6. *Ibid.*, pp 93-104.
7. *Ibid.*, pp 1-15.

15

Maintaining the Base of Chemical Research: A Response to "Challenges and Visions: Chemical Research—2000 and Beyond"

W. O. Baker

Retired Chairman of the Board
Bell Labs Innovations; Lucent Technologies
Home address: Spring Valley Rd., Morristown, NJ 07960

Human affairs have been influenced by perceptions of physics, astronomy and cosmology that have governed thought patterns through the course of the history of civilization. This commentary proposes that modern chemistry, which now defines matter in terms of its molecular and atomic existence and has the potential to design on demand the entities that will make our society function, is ready to become an integral component of the core of human knowledge. At this time, when dramatic educational reforms are being discussed for the next century, there is an exciting opportunity to devise ways to teach children from their earliest learning experiences how to deal with and what to expect from the matter of life and living. This learning must be extended to the approximately 52,600,000 students who will be enrolled at the primary and secondary levels of the nation's schools. The infusion of chemistry into our cultural awareness will further energize the research endeavor and the industrial infrastructure.

Modern chemistry is enabling a new stage in human affairs. This can come from implanting into daily doings an intuitive sense of the aggregation of matter into solids and liquids. Ongoing physical, ultrananoscopic knowledge of single atoms and molecules and how they react, self-assemble and transform can begin in infancy, as the senses of gravity, solar and moon vision, feelings of motion and energy have become "instinctive" from post-natal experience and parental coaching. Cultural rules of astronomy and physics have gained hugely from this, in the course of civilization.

So such a new climax in the sensing of matter, of all the tangible things that people live with, requires a skillful new distribution of knowledge beyond realization of the phase rule. People, especially children, could begin to live with the idea that everything they handle is an aggregate of chemicals capable of rational identification and description. Such a culture, if achieved as implied, would establish a sense of the tangible, tactile environment fostering a sustainable human existence on earth.

A New Role for Modern Chemistry in Human Affairs

Modern chemistry is ready for a new role in human affairs, in expanding ways in which science does join in the thoughts and actions of most people on earth. Such a move, based on emerging chemistry and educational culture, could in turn energize research, through its challenge and vision. This is affirmed by the influence of earlier science initiatives, that have already become part of human culture, in the deep sense of doings and feelings in daily life. These concepts have been largely in physics and cosmology, such as the realization from Galileo that the earth revolves around the sun; from others that it is not flat; that gravity from Newton is universal; from Harvey about the circulation of blood in the body and increasingly a basic intuitive feeling from the Greeks and J. J. Thomson that atomic nuclei are prevalent and have something to do with energy and life in the cosmos. Now stirred by this New York ACS Section presentation, one is emboldened to say that in an oncoming century chemistry can move beyond its great traditions of composition and reactions into a continuum of knowledge and instinct and experience comparable to these other patterns in physics, astronomy and cosmology.

Fostering a Populistic Understanding of Matter and its Transformation. The reason is that with present momentum of chemical research and particularly the validation of molecular hypothesis by direct visualization from electron microscopy, atomic force microscopy and many derivatives, the success of materials science and engineering implies a role of new extent in people's thinking and acting about matter. By this is meant that everything we touch and handle could be recognized as an aggregate of chemical entities and then subject to the rational treatment and understanding. This leads to a place in the instincts of human behavior and expectations far simpler and more primitive than the elaboration of chemical formulas and properties, as elegant and historic as that may be. Rather it appears that if the present progress in chemistry expands into detailed understanding of tactile properties of aggregates, namely all the stuff that human beings live with, there could be injected into learning virtually from infancy how to deal with and what to expect from the matter of life and living. Indeed many aspects of colloid chemistry, of high scientific and technical style a century ago, had a brief era of seeking this pathway, involving such findings that after all high polymers are molecules, did sharpen the vision. But what is possible now requires vastly more simply extended research and development in chemistry if it is to begin this populistic understanding of matter and its transformations which could be built into human sensing.

The way will be long and hard, for it will have to include subtle, often obscure, chemical systems research and engineering. Some kind of popular study in a vast variety of work, indeed of existence, far beyond anything like "science literacy," is required. However, this is not so far from what is envisioned in the follow-on to "A Nation at Risk" such as Project 2061 of the AAAS. Namely, it seems that from childhood onward ways of thinking and acting about matter could be provided if chemistry maintains its momentum of ever greater knowledge about aggregates.

Thereby the basis for common features of solids, crystals, metals, minerals, fibers, liquids, cells and living things could give the person from birth onward a sense of reason and participation in why things feel solid, break, or are hard, or are sticky or melt or are light or heavy. "Science for All Americans" hints that such learning is now conceivable.

Especially as technology and industry pervade the planet, and design becomes the basis for invention and innovation, modern chemistry will have had to relate consistently to every form of matter. Thus, in every kind of automated manufacturing and robotics, to every kind of agriculture and aerospace, modern chemistry will have had to understand and simplify at new levels the nature of aggregates and substances. Thus a reasonable complementarity to the instinctive exercise of physics and astronomy and cosmology, and eventually of some aspects of biology, might be achieved.

The times do seem right for this daunting endeavor, as indeed is so elegantly implied in the superbly wise expressions of Professor Gerald Holton and including in the invaluable 40th anniversary issue of the journal *Daedalus*.(1) Pragmatically and socio-economically the enterprise is also appealing, even if brain-wracking. This is where the chemical industry in particular will need to develop concepts and connections like those which will help it sense that appropriate chemistry of silicon dioxide, one of the ancient and classic materials known by humankind, can be developed to transport through a single fiber 400 gigabits of photonic pulses per second. This extravaganza of light transparency is leading currently to chemically and materially assembled groups of 8 fibers able to transmit undistorted 3.2 terabits (3.2 trillion bits) of information in voice, video, data, etc. per second. This is equivalent to 90,000 volumes of an encyclopedia being moved worldwide (if needed!) in one second. Already the demand is up. And that is for just one product even though in the Information Age it is a software component. As such, this software/hardware capacity is part of the third largest industry in America and may be part of the top economic products of the planet.

Linking New Chemical Research with a Dramatic Reform of Learning

This challenge of linking new chemical research with a deeper culture has, of course, long been considered by the chemical industry particularly, with many appealing themes - "better things for better living through chemistry" - and is a powerful exercise of advertising in all media for both marketing and public relations. Yet what remains to be done is to couple this chemistry of aggregates, of tactile experience with matter, to the dramatic reform of learning. This now seems to be required in response to "A Nation at Risk," beginning in infancy or early childhood and carrying through systematically in the whole of primary and secondary schooling. The culture of science being implied, ranges from knowledge of chemical composition down to 10^{14} and 10^{16} single atom or molecules or ion species per unit volume. The realms of crystal perfection, dislocations and vacancies, through the fluctuations and statistics of liquids or some post-aggregate forms, will ultimately be connected with the vast array of products of the chemical, metallurgical and other materials industries. The extensive field experiences in Project 2061 and groups such as The National Center for Improving Science Education indicate that the depth of understanding approaches what modern chemistry is now providing about new materials teaching can be devised and practiced. It can embody learning and behavior particularly that approaches that of other cultural qualities such as a sense of energy, position, environ and the like. While these things are presumably anthropomorphically less intellectual than the chemistry of solid matter, it seems possible that there is a middle ground. Effects like self-assembly, fracture, deformation, softening, etc. can be subject to new levels of understanding and of human practice.

Infusing Chemistry in the Prescience of Humankind

At least advantage should be taken of the present condition that the scientific course of chemistry is so strong and successful in pursuit of its own internal objectives. This is reflected in such successive reports as the various "Opportunities in Chemistry" (most recently the 1985 survey of the National Academy of Sciences-National Research Council, "The Pimental Report" and of its industrial embodiment as in the American Chemical Society and affiliates "Technology Vision 2020, the U.S. Chemical Industry"). So something beyond these currently fabulously successful research-engineering missions should be undertaken. The current *Daedalus* issue indeed demonstrates that the task is awesome, yet our present experiences urge us onward. As pointed out by Gordon Moore in connection with the 50th anniversary of the transistor invention, semiconductor circuit chips (primarily with transistors) now comprise the largest number of discrete products ever manufactured. Yet these central elements are especially a result of ingenious chemical controls. DNA's presence and function in all living matter was discovered by McCarty less than 50 years ago, but, now in various centers such as Professor deLange's at The Rockefeller University, work on Telomeres is indicating how they govern the lifetime of cells and thus cellular biochemical aggregates. A combination of academic and industrial researchers are reporting on mechanics of living blood vessels.(2) All these diverse studies accent the scope of the chemical aggregates we are recognizing. American industrial research and development funding has risen to more than $125 billions a year. This, too, is involving thousands of applications of chemical aggregates whose eventual integration into globally-evolving ways of living could be organized and expressed. In the first years of the next century, some 52,600,000 students will be enrolled at the primary/secondary school levels. It is suggested that the infusion of chemistry through new levels of ultimately simplifying and validating chemistry of aggregates could be a starter in this population for a timely extension of the culture that is implicit in the prescience of humankind, acting on its earthly abode.

Literature Cited

1. *Daedalus, Journal of the American Academy of Arts and Sciences,* **1998**, *127, No. 1,* p *V.*
2. Chou, J., Fung, Y.C. *Proceedings of the National Academy of Sciences,* **1997**, *94,* p *14255.*

16

Challenges and Visions:
Chemical Research—2000 and Beyond

Panel Discussion

The following are selected comments reproduced from audio tapes of the symposium, grouped in broad categories. They reflect the concerns of the panelists and participants about the need to achieve support for basic research as essential to the quality of life in the 21st Century. Issues stressed included educating policy makers and the public about science, standards of education in our schools, and improving the scope and quantity of media coverage about chemistry. The role of the government in establishing science policy and in fostering support for basic research as well as the use of tax incentives to encourage private investment in research were explored. The discussion emphasized the problems that will ensue from population increases and the critical need for finding new sources of energy. The complexities of geopolitical issues as they impact on science policy emerged in comments about the Kyoto Conference.

Science Research and Government Policy

George Olah: The twentieth century produced enormous progress. If we go back just another century, we are at the time of the Industrial Revolution. The Industrial Revolution really liberated man from slave labor, and helped to expand all aspects of our life. With all the progress of the twentieth century we, however, also introduced new problems affecting a sustainable and environmentally adaptable future. Men for survival need not only food, water, shelter, clothing, etc. but also energy. Our most widely used fossil fuel sources are limited. When politicians talk about doing something about this problem, particularly our oil supplies, they occasionally feel that something like adding a few cents to the tax to the price of gasoline may bring a solution, but even this is generally rapidly forgotten. The simple facts are—and I want to make my point strongly—that our fossil fuel resources are limited and by the end of the coming century much of our resources (maybe except coal) will be substantially exhausted. Thus, if we proceed with "business as usual", we are going to use up our future. We are indeed facing a very serious problem. I am confident we will find solutions, but these solutions will be difficult, costly and the adjustment painful. If we really do care about the future of our grandchildren and their children, we just can't sit here and do nothing. It is a global problem. We, some 270 million Americans, have created a better life for ourselves than most of the rest of the world but we need to assure that this can continue in the future. The world is rapidly becoming a global entity and we need to worry about the essential ingredients of life and a decent standard of living, not only for us but for all mankind. It is the time now to start doing something about the world's energy and environmental problems. If we would spend 1% of what we are spending on national defense to attack the long range problem of our future energy needs and environmental problems, we could be assured that when the time comes, we will have the solutions.

Ladies and gentlemen, inevitably the time will come, and I don't say tomorrow or the day after tomorrow, but in the midst of the next century when prices of essential fossil fuels will reach levels which nobody even dares to think about today. Our standard of living will be greatly affected. Still we are sitting here and are doing very little. What we know today is that the only readily available and clean energy will come from atomic energy, however, made safer and having solved the problem of radioactive waste storage and disposal. If we were able to build the atomic bombs, then, given support and priority, we certainly should be able to solve these problems and utilize atomic energy for the benefit of mankind.

Doing research including fundamental, basic studies, I believe, is essential for our future, and I hope our country will realize its continuing importance.

Bob Walker: I agree with George Olah's point. But understand that if we took 1% of the defense budget, that's 2 and 1/2 billion dollars, that's is less than what our nation spends today. So, we are not talking about huge amount of budget there. We did reduce defense budget by about $30 billion. We've taken it down to $250 billion now. Did science benefit enormously from that? Not really. The fact is that the money got diverted into all other kinds of priorities that people fully endorsed. I met an Amishman when I was first getting involved in politics and he told me that for every kind of complex problem there's a simple easy answer and it's wrong. What I found was that in government, that's exactly right. What you are dealing with all the time is an enormous number of complexities that do demand some very complex solutions. I think that that is exactly the point you make. We have people right now who have to deal with a lot of very complex issues. I would suggest to you that is high time we begin to think about what will new energy begin to look like, because fossil fuels are not a potential toward a good part of the end of the next century. So we ought to be moving toward a hydrogen economy or something like that that begins to review the problem. But getting people to focus on that when their lives are based upon a lot of petroleum and all the things that go with that becomes a complex political problem, particularly when people are not well educated because none of this is very well covered in the press. And what is happening from my standpoint in some of the press coverage? You got a good presentation here today on that. How has science reacted to this? Well, look at fundamental breakthroughs. Anyone who comes up with even the smallest breakthrough gets a PR agent and he finds it on the pages of USA Today as a fundamental breakthrough. Then, someone in the public policy area takes this "fundamental breakthrough", which isn't one really at all and gives it to a public policy advisor and public policy goes way in the wrong direction. So there has to be some reestablishing of what we really mean by even fundamental breakthroughs.

William Lipscomb: I recall that there was for years a President's Science Advisory Committee (PSAC), which was terminated several years ago. Even as recently as 1995, a House-Senate conference committee terminated the Office of Science and Technology Assessment (OTA). Recent Congresses have decreased the staff of congressional committees by about one-third, resulting, in part, in replacement of staff that have science degrees by those having politically related degrees. Even the review committees of the National Academy of Sciences are under pressure to include members from industries that are involved in areas that they study. Thus, there is a pattern of our leaders and elected bodies to reduce even further the inadequate supply of unbiased information available for important decisions and legislation about issues involving science. We are also seeing a trend toward reduction of detailed policy advice and of hearings. (ed. Dr. Lipscomb subsequently provided the following reference for this comment: *The Journal of NIH Research*, **1995**, *7*, pp 25-27.)

Bob Walker: Yes. That's absolutely right. And understand, if all this information is put forth on the Web, that's bad news for Congress, at least as it is presently constituted. Because the place where they get their information is off the general media and most of them are not even very familiar with how to use the Web. I spoke to some congressmen the other day and suggested that they do something and bookmark their information on their PC's and I looked around at their blank faces. They had no idea what I was talking about. Because a very high percent of Congress is not even computer literate—and I mean that in the most fundamental way. So, there is a real dearth of understanding.

Ronald Breslow: Let me say something about whether there is less scientific information available to Congress. I think the answer is yes. I think that the American Chemical Society is doing something about it. Since many of you are members, you should know what the ACS is doing with your dues money. We are really starting a program which is moving fairly well and is producing science and technology advisory committees for key congress people. We have done that for several people already.

The Research and Experimentation Tax Credit as an Incentive towards Furthering Basic Research

Brandon Wiers: I have a question that is intended to address the subject of the creative use of the tax code for encouraging investments in basic research in the universities which was mentioned by Representative Walker. I want to say, first of all, that our company (Proctor & Gamble, Co.) does support that (specifically, the Research and Experimentation Tax Credit) and makes use of that on a regular basis annually. However, other company's don't find it quite the incentive that would encourage its more widespread use. This was the subject considered by an initiative that our company presented to the Council for Chemical Research on that very subject this past year. Subsequent to the discussion of it, it was extended a year. And, of course, we would recommend making it permanent rather than annually extending it just one year at a time. But, to achieve the objective that you recommended, I think, that of encouraging research investments at the university levels, I would indicate that first of all there already is such a provision. But it is only at a level of 20% credit. My suggestion would be an increase in that level of credit which has in fact already been done in the State of California where it goes above the 20% credit. The other thought is that it could be extended to federal lab investments in research which are now treated as contract labs but not as basic research agencies. That could be changed. And those duel inducements with the higher level of credit could serve the purpose of encouraging industrial private sector investment in basic research to a greater degree. Those are the suggestions. I wonder what your comments may be?

Bob Walker: I think that there is a whole range of potential acts, changes that could be made encouraging basic research. It depends on where we go in terms of simplifying the tax code. Your suggestion contains excellent ideas. We need to think about developing a tax policy. Maybe what you do is allow companies tax free profits on any amount of money that they spend on basic research. In other words, you spend money on basic research, we will allow you profit tax free up to the level spent. My guess is that might encourage companies to look a little more favorably on doing some research. I think there is whole host of fairly doable ways we might pursue that would bring about a tax code that would in an overall sense bring about more research.

Brandon Wiers: The problem is that corporate welfare raises its ugly head when you talk in terms like that to academicians typically and also to federal labs many times.

Bob Walker: But, understand, the attitude of policy makers is that they have not bought into the idea of corporate welfare as a tax item. So, corporate welfare that's on the appropriations side, spending side. Yes, if you have a company getting benefit from an actual spending stream, that becomes a problem, but on the tax side, a lot of people say, look its their money, that's a different criterion.

The Kyoto Conference as a Geopolitical Issue

George Olah: We discussed some of the problems which are real. There is going to be in six weeks a conference in Kyoto to discuss an international treaty on green house gases affecting global climate. I don't think it will be able to offer any real solutions and may cost the US, depending on estimates, hundreds of billions of dollars, but still it is the first step to focus world-wide attention on the problem. I hope the scope of Kyoto will be extended to also include efforts to find new solutions, such as recycling of carbon dioxide into useful fuels.

Bob Walker: The fact is that that is an excellent example of the point I am trying to make. First of all, we have to look at these in a national context and look at the overall impact on our nation. And, secondly, we have to look at global issues. What you have in that whole process is a geopolitical issue, because we are being gamed by the

Europeans, who know that, whatever the formula is, they will find the ways and means to use old Eastern European standards to meet their criteria on it. The Japanese suddenly awoke to this the other day and decided that they didn't want to be part of the European derived formula. A science issue is involved beyond that. But it is fundamentally a political issue, a geopolitical issue, and this is the basis of what we are going to be deciding in Kyoto.

Dudley Herschbach: However, we should recognize that, for example, catalytic converters might not have come about without political pressure and lawsuits generated by environmentalists. In some cases, such pressures may be premature or overwrought when assessed in terms of the state of scientific knowledge. Even then they may spur worthy efforts. In our democracy, many things work backwards. For instance, despite laments about it, in effect our courts are often forced to legislate. That happens because we are a litigious society and major questions land in the courts long before the political battles have subsided enough to produce legislation. Likewise, the Kyoto conference deals with a genuine, important issue for which as yet there is no consensus on a "real world" solution. At least Kyoto will serve to push in the direction of trying to do something. The issue belongs on our National Agenda and might not get there otherwise.

Bob Walker: The big push can be towards moving a tremendous lot of industry off shore and giving the Europeans the opportunity to sustain their economies at our expense. There certainly will be big action, it may not be something which is in our national interest.

Multinational Corporations and the World Population Growth

Robert Curl: I would like to make a comment. I think that one hope that the world has as far as the population problem is concerned is to be found in the multinational corporations. These corporations build factories where labor costs are low and raw materials are conveniently accessible. This has the effect of developing the underdeveloped countries chosen for factories. It is quite clear that developed countries have a low population growth rate. Through such a process, the planet's population growth rate can diminish. This is my dream scenario.

Bob Walker: It may make a difference. But the central point seems to be that what often happens is that because we don't like certain policies in those countries, we impose sanctions on our companies, thereby, impeding the development. Those policies create a host of problems. Not out of malice but simply because of unintended consequences in policy decisions.

The National Education Standards: Pro's and Con's

Jean Delfiner: (High school chemistry teacher, a councilor of the New York Section, Chair of the Westchester Chemical Section, and co-chair of the High School Teachers Topical Group) The key (to the problem) seems to be world population. But, underlying that is scientific literacy. There is something that we can do now in this country. How many of you are familiar with the National Education standards, especially the standard in chemistry? The High School Chemistry Teachers by and large feel that the standard that is out there undermines the existing courses. It is not strong enough. We are really worried that we are going to have a "dumbing down" of chemistry education, the courses will become obsolete. Chemistry educators are an endangered species. The states are now going to set their standards based on the national standards and the subsequent assessments will be based on the national standard. What can the ACS do? What can industry do to see that it is changed, that chemistry gets its right place in the sun, so that we don't hear comments such as "I never understood chemistry, I don't know anything about science." You can't get journalists to write intelligently about science, let alone chemistry. What can we do? What can the ACS do? That's my question.

Dudley Herschbach: One thing the ACS can do is provide a registry of chemists, indexed in some useful fashion, ready and willing to talk to reporters and provide reliable information and perspective.

Ronald Breslow: We have a news service. They have a list of people that they know they can call.

Bob Walker: I would simply say to you that changing the standard is probably not going to do you much of anything. Any time you start to develop national standards, there always develops the lowest common denominator. That is why anybody involved in education ought to be suspicious any time somebody in Washington suggests that there is some national standard that is to be developed. There is always going to be "dumbing down", and it's always going to be the lowest common denominator.

Dudley Herschbach: Actually, the National Standards for Science Education were not developed by "somebody in Washington." The project involved a host of scientists and educators from all over the country. It was conducted by private, nongovernmental agencies, led by the National Academy of Sciences. Unfortunately, it is now politically fashionable to impugn anything containing the word "National." The need for such standards becomes clear when you compare K-12 science education here with that in other countries. But I hope that the old prideful American spirit will spur many schools and even a few states to exceed the National Standards, by a lot.

For university science faculty, I'd like to advocate a simple step that would encourage the natural enthusiasm for teaching and expounding that many graduate students have. We could foster a tradition of including in the Ph.D. thesis a chapter in which the candidate describes an educational initiative, and/or another that explains the work in terms understandable to, say, a member of congress. Such chapters should I think be optional rather than required, but encouraged by some reward mechanism. This would help to broaden the student's perspective and convey that we want a Ph.D. to be not a technician but an architect.

George Olah: I would like to bring up an additional point concerning the fundamental role of education. The best thing we can do for the future of our young people and the future of our nation is to provide a very good education. It's essential. As a personal reflection, I was born and raised in Hungry, a small and, at that time, in many ways a rather underdeveloped country in Europe which, however, somehow produced a number of fairly good scientists, technologists, etc. I believe the reason was that there were a number of good high schools based on the German high school systems, and students were stimulated and interested in science. I think that schools can have an enormous influence.

Chemistry and the Media: The Science Fair as a Human Interest Story

Question posed by an unidentified student: My question is primarily directed to Mr. Service. I read the media on science while I was growing up. I am concerned about the image of science because between grammar school and high school there is something critical that happens that deters a lot of students from fields like chemistry and physics, and in turn, science loses a lot of enthusiasm, and loses attention, and thus loses a lot of competent youth as well. Now I and a lot of my peers are concerned about the stereotypes of science that we see in the public image and the pop culture. Have you seen any change?

Robert Service: None that I can think of. However, one solution may be to find ways of teaching about chemistry through people. What is needed are stories about people involved in chemical research. The human interest stories do get published and are read. This approach might galvanize people.

Robert Lichter (The Camille and Henry Dreyfus Foundation): I want to pick up on Mr. Service's comments about telling stories about people. Dudley Herschbach earlier this morning asked "How many people here know about the international science and engineering fair?" A few people raised their hands. He did not ask "How many people heard or read about it in the newspapers?" Probably fewer knew or heard about it from the newspapers. These things aren't covered. Actually that local newspaper that Ronald Breslow gets delivered and that I sometimes get delivered did have something on page D-16 about it. Nobel prizes in chemistry and physics were on page A-16. I've asked this question of reporters on the newspapers: "Why, for example, in New York City did the science fair that used to be held in New York not make into the press?" We were talking about young people. The real people stories—as some of you so eloquently noted—and yet, it just doesn't make it into the press. The answer was: "That's not a science issue, that's the Metro Section." O.K. To me the solution is simple. You take two tin cans, a piece of string and run them between the Science Section and the Metro Section, and get them to talk to each other. I appreciate the structure that characterizes the way the press works and to some extent I even sort of understand it. But, I'm concerned that it becomes an excuse and not a reason. I really would like to see those who are playing significant roles in deciding what gets covered to really take a very hard look at those very fundamental issues and even convert them creatively (using some of the thinking that Mr. Walker is suggesting) into how that might affect the bottom line.

Dudley Herschbach: Overall, science education and literacy in the United States are feeble. But there are some strengths to build on. Among them are the Westinghouse Science Talent Search and the Intel International Science and Engineering Fair. Anyone who attends these events or serves as a judge will become a lot more optimistic about our future. The high school kids who enter are doing the real thing; on their own initiative they take ownership of some project. In the course of developing it, and exhibiting it, often at a series of fairs, they arouse the interest of friends and family and lots of curious neighbors.

Both the Talent Search and the International Fair are conducted by a small nonprofit outfit, Science Service. It also publishes *Science Néws*, an eight-page weekly, written for laymen, that provides an excellent survey of what's happening in all fields of science. For 30 years, Glenn Seaborg chaired the Board of Science Service, and a few years ago he recruited me as his successor. Science Service now hopes to get *Science News* into every high school in the country and to further enhance the Talent Search and Fair.

Last year Intel undertook to sponsor the International Fair, which has existed for nearly 50 years but is, as yet, much less well known than the Westinghouse Talent Search. The Intel Fair is held every May in a different city. In 1997 it was Louisville, in 1998 it will be Fort Worth, the week of May 12. The Fair has grown greatly in recent years. At Fort Worth, there'll be projects displayed by1200 kids from 50 countries, over 90% from the U.S. Remarkably, those kids are all winners of hundreds of local, state, and regional fairs in which about a million others took part! The Fair also involves a thousand volunteer judges and hundreds of volunteers involved in running it. In Louisville, about 80 four-year scholarships were given as prizes.

Considering all the friends, relatives, and teachers of the kids entering the preliminary fairs, all told there must be several million people with links to the Westinghouse Search and/or the Intel Fair. The kids displaying and explaining their projects are fine ambassadors for science. I wish the major media would pay more attention, particularly to the Intel Fair. I'd like to see TV news programs include, just as regularly as the weather report, a one or two minute episode featuring a student presenting an engaging and instructive project. That would surely attract a devoted viewership, since so many other kids, parents, and teachers would want to tune in. A year's supply of such segments could be taped, with unusual efficiency, at the annual Westinghouse and Intel events.

Here's a nice example of a striking project, actually one I didn't get to see but heard about. A student took 30 mice, ran them through a maze, and found on average they took 30 minutes to emerge. Then for 8 hours he had 10 of the mice listen to classical music, 10 listen to rock music, and 10 serve as a control group which listened only to each other. He found the control group still needed 5 minutes to navigate the maze. But the mice edified by classical music did it on average in only a minute and a half, while those stupefied by rock music required 20 minutes! That ought to be front page news.

Index